Interactive College Algebra
A Web-Based Course

Student Guide

Davida Fischman
Terry Hallett
Dan Rinne
Peter Williams

www.keycollege.com

Davida Fischman, Terry Hallett, Dan Rinne, Peter Williams
California State University at San Bernardino
San Bernardino, California

Key College Publishing was founded in 1999 as a division of Key Curriculum Press® in cooperation with Springer-Verlag New York, Inc. We publish innovative texts and courseware for the undergraduate curriculum in mathematics and statistics as well as mathematics and statistics education. For more information, visit us at *www.keycollege.com*.

Key College Publishing
1150 65th Street
Emeryville, CA 94608
(510) 595-7000
infokeycollege.com
www.keycollege.com

© 2005 by Key College Publishing. All rights reserved.
Published by Key College Publishing, an imprint of Key Curriculum Press.

The work consists of a printed book and CD-ROM packaged with the book, both of which are protected by federal copyright law and international treaty. The book may not be translated or copied in whole or in part without the permission of the publisher, except for brief excerpts in connection with reviews or scholarly analysis. Use in connection with any form of information storage and retrieval, electronic adaptation, computer software, or by similar or dissimilar methodology now known or hereafter developed is forbidden.

All other registered trademarks and trademarks in this book are the property of their respective holders.

Interactive College Algebra CD-ROM

Key College Publishing guarantees that the CD-ROM that accompanies this book is free of defects in materials and workmanship. A defective disk will be replaced free of charge if returned within 90 days of the purchase date. After 90 days, there is a $10.00 replacement fee.

Development Editor: Elizabeth McCue
Production Director: Diana Jean Ray
Production Project Managers: Michele Julien, Ken Wischmeyer
Copyeditors: Jessica Holden, Erin Milnes
Proofreader: Erin Milnes
Art and Design Coordinator: Kavitha Becker
Cover Designers: Keith Nyugen, Mary Shu
Composition: Interactive Composition Corporation
Printer: RR Donnelley

Editorial Director: Richard Bonacci
General Manager: Mike Simpson
Publisher: Steven Rasmussen

Printed in the United States of America
10 9 8 7 6 5 4 3 2 1   07 06 05 04

ISBN: 1-931914-23-0

In memory of my mother, Vera Wynn Williams, who always gave me her support and encouragement – Peter.

To my Mother, Joan Collar, for her unfailing support – Terry.

To Cindy, Heather, and Ewen – Dan.

With love to my ever-patient and always supportive husband, Shlomi – Davida.

About the Authors

Davida Fischman received her Ph.D. in 1989 from Ben Gurion University, Israel. Her research interests are in Hopf Algebras. Since 1994 Davida has been a professor at California State University, San Bernardino. Prior to that she worked as a visiting professor at the University of Southern California and at the Weizmann Institute, Israel. Davida is the coordinator of the Master of Arts in Teaching Mathematics at CSUSB. Davida has always been very involved with mathematics education and the use of computers in the classroom.

Terry Hallett received her Ph.D. from the University of London, England, in 1961. Terry has been a professor at California State University, San Bernardino, since 1981. Prior to that she held positions at Royal Holloway College, University of London, University of Nevada, Reno, and California State University, Northridge. Terry has been the graduate coordinator for the Master of Arts in Mathematics since its inception. She has served as Developmental Math Coordinator at CSUSB for the past three years and served as Associate Dean for the College of Natural Sciences from 1987 to 1993. Terry encourages increased use of technology and web-based activities in the classroom.

Dan Rinne received his Ph.D. in 1979 from the University of California, Santa Barbara. Dan's research interests are in real analysis. He has been a professor at California State University, San Bernardino, since 1982. Prior to that he held teaching positions at Metro State College in Denver, Colorado, and Feng Jia University in Taichong, Taiwan. Dan has been involved in the use of computers in the classroom for several years.

Peter Williams received his Ph.D. in 1983 from the University of St. Andrews, Scotland. His research interests are in combinatorial and computational group theory. Peter joined California State University, San Bernardino, in 1983 and has served as department chair since 1997. Peter has long had a keen interest in the use of computers in the classroom, and he has used software in a wide variety of classes.

Contents

About the Authors   v

Preface   xi

Acknowledgments   xiii

Introduction   xv
    To the Instructor xv
    To the Student xvii
    Using the Modules xix
    Typing Mathematics on the Web xxii
    Quizzes xxvi

1 Coordinate Plane   1
    1.1   Structure of the Plane 1
    1.2   Distance and Midpoint 7
    1.3   Lines 13
    1.4   Circles 25
    1.5   Complex Numbers 31
    1.6   Broken Wheel Problem 37

2 Functions and Graphing   39
    2.1   Functions 39
    2.2   Graphing Techniques 49
    2.3   Quadratic Functions 61
    2.4   Maximizing the Area of a Rectangle 67
    2.5   Graphing Quadratics 71

3 Operations and Inverses   75
    3.1   Operations on Functions 75
    3.2   Inverse Functions 83

4 Polynomial Functions   93
    4.1   Polynomial Functions 93
    4.2   Finding Zeros 101
    4.3   Division 102
    4.4   Zeros and Graphing 107
    4.5   Synthetic Division 113
    4.6   Zeros and Factoring 121

- 4.7 Descartes' Rule of Signs 125
- 4.8 Bounds for Zeros 129
- 4.9 Rational Zeros 135
- 4.10 Complex Zeros and Coefficients 141
- 4.11 Approximating Zeros 143
- 4.12 Factoring and Graphing 144

5 Rational Functions — 149
- 5.1 Rational Functions 149
- 5.2 Domain and x-intercepts 155
- 5.3 Sign Determination 159
- 5.4 Vertical Asymptotes 163
- 5.5 Horizontal Asymptotes 169
- 5.6 Graphing Rational Functions 175

6 Exponential Functions — 181
- 6.1 Exponential Functions 181
- 6.2 Graphing 189

7 Logarithmic Functions — 195
- 7.1 Introduction to Logarithms 195
- 7.2 Base 10 and Base e 201
- 7.3 Inverses 209
- 7.4 Properties of Logarithms 215
- 7.5 Logarithmic and Exponential Equations 219

8 Systems of Equations — 227
- 8.1 Systems of Linear Equations 227
- 8.2 Equivalent Systems 233
- 8.3 Infinitely Many Solutions 236
- 8.4 Solving Systems of Equations 241
- 8.5 Mixture Problems 249
- 8.6 Parabola from Three Points 253

9 Matrices — 257
- 9.1 Matrices 257
- 9.2 Systems of Equations and Matrices 265
- 9.3 Determinants 273
- 9.4 Cramer's Rule 279
- 9.5 Inverse Matrices 283
- 9.6 Systems of Equations and Inverse Matrices 289

10 Sequences — 293
- 10.1 Sequences 293
- 10.2 Arithmetic Progressions (AP) 301

CONTENTS

    10.3 Sums of Arithmetic Progressions 307
    10.4 Geometric Progressions (GP) 311
    10.5 Sums of Geometric Progressions 317
    10.6 Repeating Decimals 323
    10.7 Amortized Loans 327

A **Completing the Square**   **331**

B **Broken Wheel Example**   **333**

C **Synthetic Division Example**   **335**

D **Answers to Selected Exercises**   **337**

Index   **347**

Preface

A college algebra course is typically the course of choice for most students that need to satisfy a mathematics general education requirement. In teaching college algebra over many years in a traditional setting, we observed and were concerned about lack of student participation and the high attrition and failure rates. We were motivated to incorporate computers into the classroom based on other universities' success stories about incorporating technology into the classroom. Although some computer software was available, we were not able to find anything that fit our ideas on how to incorporate technology into our college algebra courses. We wanted something that would increase student participation and appeal to the students so they would want to engage actively in mathematics. We were also looking for software that would operate on a variety of platforms. Using an Internet browser combined with JavaScript and Java seemed to be ideal for meeting our goals. We began creating the *Interactive College Algebra* modules in 1998 with the aid of a grant from our campus. We developed applets to demonstrate the concepts students would encounter in a college algebra course and that would allow students to practice what they had learned. The first chance we had to try out the material was in Spring 1999. The group of students who worked with this new material were very pleased with it. They could work out problems time and time again in a relaxed, nonthreatening environment and were able to visualize topics, such as translating a graph, as well as participate in moving the graph. The software gave them a more hands-on approach.

Our observations from the pilot project were very encouraging. We noted increased student participation, a closer sense of community within the group of students, and a lower attrition rate. Students seemed appreciative of the interactive nature of the material as well as the practice quizzes.

The pilot project was funded through a grant from the Teaching Resource Center at California State University, San Bernardino. Through a grant from the National Science Foundation, we were able to expand the project to cover material for a typical college algebra course. We have continued to make changes to the material based upon suggestions from both students and instructors.

In contrast to other technology-based mathematics courses that use technology as a medium for presenting material to students, *Interactive College Algebra* increases motivation by allowing students to interact with technology in a web-based format, providing an appealing alternative to the traditional text-based approach.

The web-based modules provide detailed explanations of topics; the Java applets demonstrate concepts and provide students with numerous practice problems. Students therefore can review algebraic ideas and practice new skills until they have mastered them. They can also take practice quizzes to assess their progress.

The web-based format of this course also allows for advanced graphics and animation that help make abstract concepts easier to understand through visual representation. *Interactive College Algebra* is easy to use, requires no prior computer skills, and encourages students to engage actively in mathematics. The primary goal of the interactive courseware is to increase student retention of subject matter and increase success rates through changes to the curriculum and teaching strategies. We hope this unique interface brings motivation and success to your classroom.

Overview of *Interactive College Algebra* Supplements

Instructor Resources A printed *Instructor Resources* guide is available to qualified adopters. This valuable resource includes teaching tips on the most effective use of the interactive course materials and ideas for course planning based on scheduling options and topic coverage, as well as instructions and objectives for each online module. The authors provide insights into the unique challenges of assessment, particularly the advantages of using a combination of online and traditional tests. Printed solutions to selected exercises are included.

Test Check Test-generating software is also available to instructors. This valuable assessment tool offers advanced course management, performance tracking and scoring, web assessment, and a bank of test questions unique to *Interactive College Algebra*. Instructors are able to build quizzes and tests and post them to the web for students to take and submit online, or print them for students to complete in class or as homework. Additionally, when students submit tests online, this software automatically scores their multiple-choice answers, which instructors can then import to their electronic grade books.

Web Resource Center Other valuable resources for students and instructors are available at the Key College *Interactive College Algebra Web Resource Center* at **www.keycollege.com/ICA**.

For more information about *Interactive College Algebra*, please contact your Key College sales representative toll free at 888-877-7240 or visit www.keycollege.com.

Acknowledgments

The project *Interactive College Algebra: A Web-Based Course* was funded in part by grants from the Teaching Resource Center of the California State University, San Bernardino, and the National Science Foundation (*Course, Curriculum, and Laboratory Improvement—Educational Materials Development* Grant #9952821).

The authors would like to thank the faculty of the Mathematics Department at CSUSB who provided advice and suggestions on many facets of this material. A special thanks to Professors Chris Freiling and Shawnee McMurran, who tested the material in class and who provided valuable feedback to help improve the presentation of the material. We are also grateful to the editors and staff at Key College Publishing for their unflagging support of the project, their limitless help and guidance in the process of publication, and for their dedication and professionalism.

The authors and publisher wish to thank the following reviewers for their invaluable feedback throughout the development of this courseware:

- Linda Buchanan, Howard College, TX
- Janis Cimperman, St. Cloud State University, MN
- Frederick Lane, Palm Beach Community College, FL
- Kim Luna, Eastern New Mexico University, NM
- Margaret Morrow, Plattsburgh State University, NY
- Pamela Smith, Fort Lewis College, CO
- Gary Stoudt, Indiana University of Pennsylvania, PA

We also wish to thank Jodi Cotten from Westchester Community College in New York for her technical review of *Interactive College Algebra*.

Finally, we would like to thank our families for their support and encouragement throughout the many years devoted to development of these materials.

Introduction

To the Instructor

Materials

The source materials for *Interactive College Algebra: A Web-Based Course* consist of the web site, the Student Guide, and the accompanying CD-ROM. Both the Student Guide and the web site are required to effectively cover all the topics included. The CD-ROM contains all the material available on the web and can be used when a student does not have access to the Internet. The major features of each component are as follows.

Student Guide

- Complete mathematical content
- Illustrations of interactive exercises contained on the web
- Illustrations of demonstrations contained on the web
- Special instructions related to interactive exercises
- Additional written exercises on each topic

Web Site and CD-ROM

- Complete mathematical content
- Interactive exercises
- Interactive demonstrations
- Practice quizzes on each topic

Please note: Use of the material on the web site or the CD-ROM requires a web browser that is JavaScript- and Java-enabled. For example, Netscape 4.0 or higher and Microsoft Internet Explorer 4.0 or higher have these capabilities.

Course Structure
Topics
The modules cover the following topics. The flow chart shows the interdependence of the modules.

1. Coordinate Plane
2. Functions and Graphing
3. Operations and Inverses
4. Polynomial Functions
5. Rational Functions
6. Exponential Functions
7. Logarithmic Functions
8. Systems of Equations
9. Matrices
10. Sequences

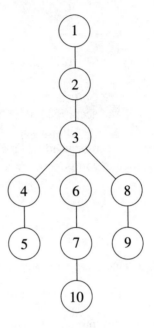

Time Schedule
The material in the first module, Coordinate Plane, should be known to most students in college algebra, and it is possible that this module could be either quickly reviewed or skipped altogether. The module on Polynomial Functions is fairly long and can be allotted more time, depending on interests, by reducing the amount of time spent on a module such as Matrices or Sequences. Again, how much time an instructor spends on each module depends on the importance of each topic in the overall course plan. In a 15- or 16-week semester, every page beyond the first module could be covered at a rate of about one page per day in a class meeting three times per week. In any setting, some module will be deemphasized or skipped if the entire first module must be covered in detail.

To the Student

Using the Student Guide

Studying the Mathematical Content

This guide can be used to study the course material without the web, as it contains a copy of the web pages as they appear on screen. Thus, the definitions, theorems, and examples can all be accessed without the web. Using the guide along with the web makes it convenient to read the mathematics while keeping an exercise visible on the screen.

Special Instructions

Special instructions are included in this guide to advise you on the interactive exercises. These instructions stand out from the body of the text, appearing like this:

This bit of text would contain instructions concerning the next exercise, and you would be directed to go there now if simultaneously using the web site.

Directions to Interactive Exercises

The interactive exercises on the web, as well as the special instructions mentioned above, are set off from the rest of the text by a mouse icon that extends into the margin and enclosing horizontal lines. An exercise will appear similar to this one, taken from the module on the Coordinate Plane:

Now try the exercise on the web page. We recommend that you do several sets of this exercise. In each set you are required to plot six given coordinates.

You will practice plotting points and translating geometric representations of a point in a plane to its x- and y-coordinates in the following activity. Click on New to get a new list of points. The arrowhead (>) will point to one of the pairs of coordinates. Click on the graph at the point that these coordinates represent. If you have clicked on the right point, you will see a red dot appear, and the arrowhead will move down one coordinate pair. If not, you will be told the coordinates of the point on which you have clicked, and you can try again. Continue this way until you finish the list. To get a new list, click on New again.

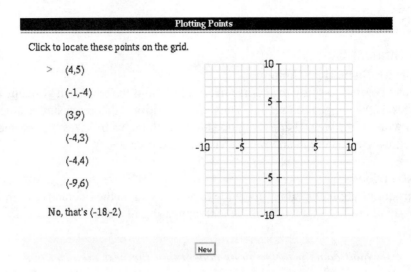

Written Exercises

At the end of each module is a set of written exercises covering the topics of that module. These can be used as practice exercises without accessing the web. Included are problems to further establish the methods of the module as well as word or application problems. You should be able to do these in addition to the interactive exercises.

Using the Modules

Starting the Modules

If you are visiting *Interactive College Algebra* online for the first time, you will need to register for the course with Key College Publishing at *www.keycollege.com/online*. There you will be prompted to enter the single-use registration access code printed on the card in the back of your Student Guide. After you enter the access code, it will expire, and you will be asked to create the unique username and password that you will use to access the online material for the duration of your course. You can log on to the course material through *the Interactive College Algebra Web Resource Center* at *www.keycollege.com/ICA* or by bookmarking the course home page.

If you do not have access to the Internet, or you have a slow connection, you can run the *Interactive College Algebra* courseware from the CD-ROM included with your Student Guide. The CD contains the complete course material available on the Web, but it does not include the contents of the *Web Resource Center* or allow for online testing. To start the CD, insert it into your CD-ROM drive. If Autorun is active, your Web browser will launch, and you will be taken to the *Interactive College Algebra* home page. If the browser does not launch on its own, open the browser and open the file **D:index.html,** where "D:" is the letter of your CD-ROM drive. You will not need a password to access the CD's contents.

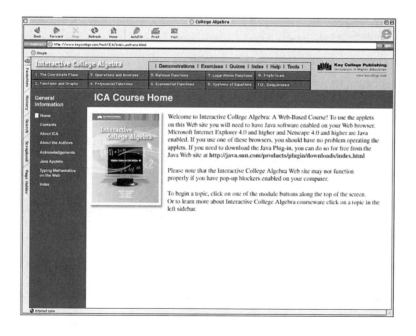

Navigating the Modules

The ten module buttons at the top of the screen bring up a list of the individual pages within that module and initially display the objectives page for the module. For example, clicking on Coordinate Plane will change the appearance to this:

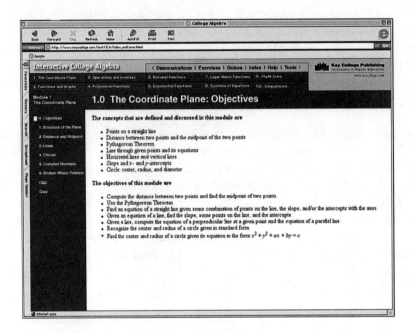

If we choose the page entitled "Circles" from the list of section heads, that page is displayed in the main frame, while the list of pages is maintained on the screen.

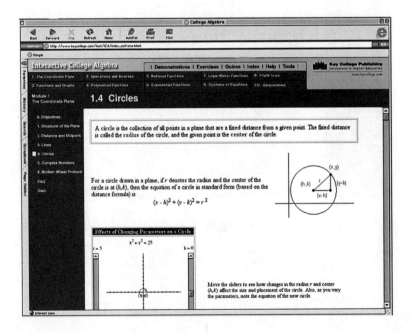

USING THE MODULES

Tools

Clicking on the ⟨Open Tools⟩ button will open a small window with three icons that access a calculator, a synthetic division tool, and a matrix manipulator. These may be opened and used when desired while studying the material.

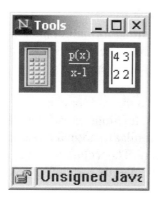

Additional Supplementary Material

The buttons in the navigation bar labeled "Demonstrations" and "Exercises" bring up lists of all interactive demonstrations and exercises in all modules. Use these lists to go to a particular activity directly if you are unsure of its exact location. In addition, each module has its own practice quiz, and the ⟨Quizzes⟩ button brings up a list of all ten quizzes. See the Quizzes section for more information. To locate a particular term or topic, an index can be found by clicking the ⟨Help⟩ button and selecting "Index" from the list of section heads.

Typing Mathematics on the Web

Displaying complex mathematical formulas on the web in any sort of familiar manner is not an easy task. Although it is now possible to display powers (such as x^2), there is no corresponding easy way to display the square root symbol and other specialized notation we commonly use and see in print.

In addition to the difficulty of presenting formulas on the web, there is also the problem of how you can enter expressions in order to complete some of the exercises and how the computer can interpret the data you have entered.

For these reasons, we have devised a method for displaying and entering formulas within the Java applets that involve exponents, fractions, or radicals. This means that most formulas will be displayed for you in a manner similar to normal typesetting in textbooks. The trickier part is entering such formulas as answers. The technique we developed is fairly simple, requiring just a little practice to get accustomed to the special key sequences we employ.

First we'll summarize some notation commonly used today by many software packages that can handle mathematical notation and functions. We will then cover the special steps we use in these web pages to handle some of the notation.

Commonly Used Symbols

Arithmetic Operators

The arithmetic operators for addition, subtraction, multiplication, division, and exponentiation are denoted by $+$, $-$, $*$, $/$, and $\hat{}$, respectively. We may use parentheses "(" and ")" to make clear our intentions as to the meaning of what we enter from the keyboard. (There is the usual implied order of operations when parentheses are omitted.) In order of priority from highest to lowest, we have

( )	Parenthesized expressions are evaluated first.
$\hat{}$	Exponents are next.
$*$, /	Multiplication and division are equal in order.
$+$, $-$	Addition and subtraction are lowest in evaluation and equal in order.

For example, $2 + 3\^4$ has the value 83 and not 625, since the $3\^4$ is evaluated first and then added to 2, yielding 83. However, $(2 + 3)\^4$ has a value of 625, since the parenthesized expression is evaluated first, giving 5, and then 5 is raised to the fourth power.

Expressions

Many times you will be asked to enter expressions in the exercises. The symbols listed above may be used, but the process is quite cumbersome. For example, the polynomial $3x^4 - 5x^2 + 7$ could be entered as $3 * x\^4 - 5 * x\^2 + 7$. It is easy to forget the $*$ and type $3x\^4 - 5x\^2 + 7$. This is one of the reasons we created our special techniques described below, in "Our System."

Functions

As there is no key on the keyboard with the square root symbol (or any other common function), one convention in common use is to write the square root as "sqrt" with its argument in parentheses. So, the square root of 2 would be written as sqrt(2). Similarly, other functions may be entered with the argument in parentheses. You are allowed to enter 11 functions by name, and these are tabulated as follows:

Notation for Common Functions

Notation	Function
sqrt	square root
abs	absolute value
sin	sine
cos	cosine
tan	tangent
asin	arcsine (inverse sine)
acos	arccosine (inverse cosine)
atan	arctangent (inverse tangent)
exp	exponential
log	(common) logarithm
ln	(natural) logarithm

Special Constants

The numbers π (approximately 3.14159) and e (approximately 2.1718) can be entered as *PI* and *e*, respectively, wherever you have to type their values.

Our System

Applets that require the typed entry of answers with no special symbols or formatting use standard Java input regions. Examples would be integers, decimal numbers, or simple fractions entered as *a/b*. Some care must be taken even with these simple types of answers.

For example, $-3/4$, -0.75, or $3/(-4)$ would all be interpreted as the same value but the expression $3/-4$ would be read as a malformed expression. The following exercise contains such a region. On the web, the background color is aqua, and the region is initially blank. Some of these will include scroll bars and some will not, depending on the space allotted and the exact nature of the answer. To enter an answer, click the mouse inside the region and begin typing. The backspace and arrow keys are functional in these regions.

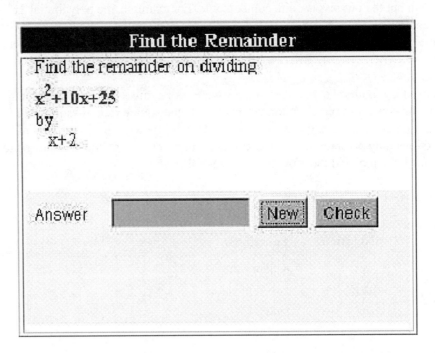

In order to allow typed answers that require special symbols or formatting, a different type of input region was created. This method allows, for example, an x-intercept of the form $(1 + \sqrt{3})/8$ to be displayed on screen in fraction form as it is being typed.

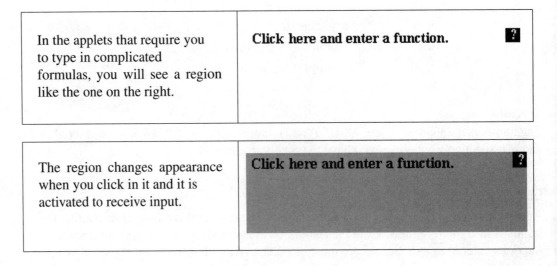

TYPING MATHEMATICS ON THE WEB

Because exponents, fractions, and square roots occur so frequently in the material we cover, we have adopted the following conventions for typing formulas in the input regions:

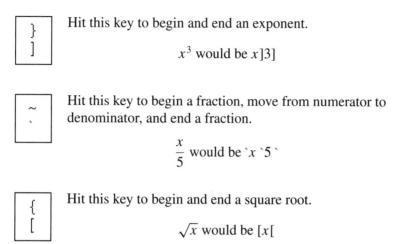

`}` `]`	Hit this key to begin and end an exponent. x^3 would be x]3]
`~` `` ` ``	Hit this key to begin a fraction, move from numerator to denominator, and end a fraction. $\frac{x}{5}$ would be `` ` ``x `` ` ``5`` ` ``
`{` `[`	Hit this key to begin and end a square root. $\sqrt{x}$ would be [x[

The "?" in the upper right corner of the region brings up a reminder of these three rules whenever the mouse is passed over the enclosing blue rectangle.

> Any of the symbols and special keys mentioned here may be used within the input region. Using these three special techniques eliminates the use of "^" for exponentiation, "/" for simple fractions, and "sqrt" for square roots. The formulas will appear on screen as they would be typeset. In addition, the "∗" is not required between coefficients and variables. For example, $3x$ can be used instead of $3 \ast x$.

The following demonstration lets you practice entering a few formulas to get used to the techniques. Try them first yourself, but if you need some help in the steps, click on ⬚Show Me to reveal the correct input. The backspace key may be used to delete unwanted characters.

Please note that if you have entered exponent, fraction, or square root mode by mistake and wish to exit it, it's best to use the backspace key. If, on the other hand, you hit the fraction key three times, you will indeed exit fraction mode, but you will have a small quotient line left. If you type "[" twice, you will exit square root mode, but you will have the square root symbol left on the screen (although the backspace will delete the "leftovers" in both cases). To avoid these situations, just use the backspace key to exit the mode.

INTERACTIVE COLLEGE ALGEBRA: A WEB-BASED COURSE © 2005 Key College Publishing

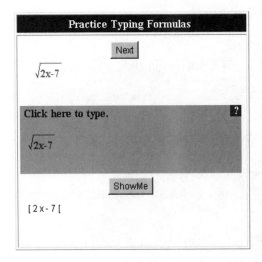

In this typing region, both $\frac{3}{-7}$ and $-\frac{3}{7}$ would be accepted and interpreted as the same value. The latter version is preferred. Fractional exponents must be entered using the "/" key. So, an expression like $2^{\frac{x}{3}}$ would be entered and would appear as $2^{x/3}$.

Quizzes

Each module is followed by a quiz on the topics of the module. Most quiz questions are multiple choice. Following is an example of how a typical quiz question appears. (This example is from the module on Functions and Graphing.)

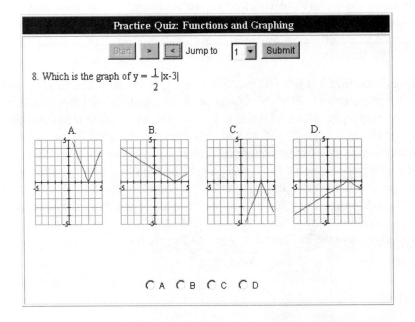

QUIZZES

After you have answered all the quiz questions, click the Submit button and your quiz will be graded, your score will appear on the screen. Correct answers will be provided for any questions missed, so they may be reworked before proceeding. The actual quiz questions are randomly generated, so don't expect to see the same questions if you take the quiz a second time. (Expect the same *type* of questions, but not identical ones.) This is why it is especially important to review any missed questions before leaving the quiz.

Module 1

Coordinate Plane

1.1 Structure of the Plane

Given a point in a plane, we can associate that point with an **ordered pair** of real numbers as follows: In the plane, draw a real number line horizontally and a real number line vertically (with increasing values going upward), so that the zeros on each line intersect. The horizontal real line is called the **x-axis**, the vertical real line is referred to as the **y-axis**, and the point of intersection is called the **origin**.

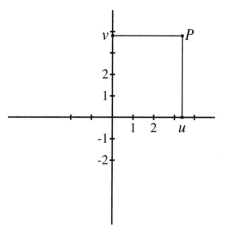

Next, construct perpendicular lines from the point to each of the axes. Suppose the perpendicular line to the x-axis meets at the real number u, and the perpendicular line to the y-axis intersects at the real number v: then we associate the point with the ordered pair (u,v). Conversely, any ordered pair of real numbers corresponds to a point in the plane. The ordered pair corresponding to a given point are called the **coordinates** of the point. Given the coordinates (u,v) of a point, we call the first number, u, the x-coordinate of the point and the second number, v, the y-coordinate of the point. When we introduce perpendicular axes into this plane in this way, we have a **Cartesian coordinate plane** (or just a **coordinate plane**).

Notice that the **coordinate axes** (the collective name for the x- and y-axes) divide the plane into four parts, called **quadrants**. We number them counterclockwise, beginning from the quadrant in which both the x- and the y-coordinates are positive, as follows:

Quadrant number	x-coordinate	y-coordinate
First quadrant	positive	positive
Second quadrant	negative	positive
Third quadrant	negative	negative
Fourth quadrant	positive	negative

Second quadrant	First quadrant
Third quadrant	Fourth quadrant

Note that points that lie on an axis do not lie in any quadrant. If a point lies on the x-axis, then its y-coordinate is 0. Likewise, a point on the y-axis has 0 as its x-coordinate. The origin has coordinates $(0,0)$.

On the right you see an example of a (Cartesian) coordinate plane with some points sketched in it.

Coordinates: Point A has x-coordinate 2 and y-coordinate 3, so we say it has coordinates $(2,3)$. Point B has x-coordinate 1 and y-coordinate -2, so we say it has coordinates $(1,-2)$.

Quadrants: Point A is in the first quadrant, and point B is in the fourth quadrant.

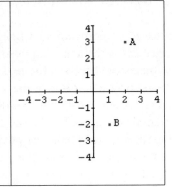

Let's move to the web now. In the following activity, you will observe the coordinates of a point as well as the quadrant in which you are located as you move the pointer around the plane. You will also be able to see the perpendicular lines that help determine the coordinates of the point. When is the x-coordinate positive? Negative? What about the y-coordinate?

In the demonstration to the right, we have drawn a pair of axes in the plane. Move the mouse over the plane to see the coordinates and the quadrant in which the pointer is located.

If you click the mouse, the demonstration will "freeze" and you will be shown the perpendicular lines from the point to the axes. Click the mouse a second time to reactivate the demonstration.

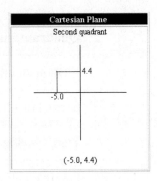

INTERACTIVE COLLEGE ALGEBRA: A WEB-BASED COURSE © 2005 Key College Publishing

1.1. STRUCTURE OF THE PLANE

If we are given the coordinates of a point, we may place that point in the plane geometrically. This is called **plotting** the point.

_____

Now try the exercise on the web page. Work out several sets of this exercise, until you can get all the points right. There are six points for you to identify in each set. In the exercise the student clicked on the point with coordinates $(-18,-2)$ instead of the correct point, $(4,5)$. In response, the computer pointed this out, and the student may now try to plot the point $(4,5)$ again.

You will practice plotting points and translating geometric representation of a point on a plane to its x- and y-coordinates in the following activity. Click on New to get a new list of points. There will be an arrowhead (>) pointing to one of the pairs of coordinates. Click on the graph at the point that these coordinates represent. If you have clicked on the right point, you will see a red dot appear, and the arrowhead will move down one coordinate pair. If not, you will be told the coordinates of the point on which you have clicked, and you can try again. Continue this way until you finish the list. To get a new list, click on New again.

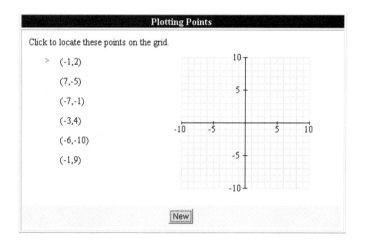

The next exercise is the reverse of the previous one in that you are given points marked in the plane and you are to identify the coordinates of the points. Being able to go back and forth between coordinates and the location of the point in the plane is a skill that you will use over and over again throughout this course and in subsequent material. In the exercise shown here, the student has correctly identified the points marked in gray on the grid ($(4,7)$, $(5,7)$, $(5,9)$, and $(3,1)$). The next step is to identify the point indicated by the black square by clicking on the appropriate coordinate pair.

In the next activity you will practice identifying the coordinates of points that are graphed. Each time you click on New , some points will appear in the graph on the left and a list of coordinate pairs will appear on the right. One of the points on the graph will be red; this is the one you are to identify. If you click on the correct pair of coordinates, a different point will become red; this is the next one to identify. If you click on the wrong pair, you can try again.

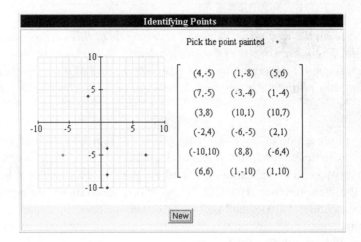

Additional Exercises

1. In which quadrant is each of the following points located?

Point	Quadrant	Point	Quadrant
$(-9,3)$		$(4.3,7)$	
$(-8,-41)$		$((-2)^5,(-4)^{-3})$	

2. If (x,y) is in the given quadrant, fill in the correct quadrant for the corresponding point.

Point	Quadrant	Point	Quadrant
(x,y)	third quadrant	$(-x,y)$	
(x,y)	second quadrant	$(-x,-y)$	
(x,y)	fourth quadrant	$(x,-y)$	
(x,y)	second quadrant	(x^2,y^2)	

3. Match the following coordinates with the points labeled A through J in the plane.

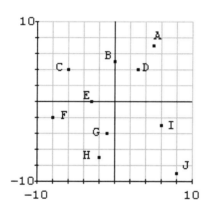

Coordinates	$(6,-3)$	$(-6,4)$	$(-3,0)$	$(3,4)$	$(-2,-7)$
Point					
Coordinates	$(-1,-4)$	$(8,-9)$	$(-8,-2)$	$(0,5)$	$(5,7)$
Point					

1.2 Distance and Midpoint

Distance

Suppose we are given two points and we want to find the distance between them. Since they are in a coordinate plane, we can use their x- and y-coordinates to find this distance.

We first consider the case of two points that lie on a line parallel to one of the axes.

- If the x-coordinates of both points are the same, then the line is parallel to the y-axis, for example, points $A(-1,2)$ and $B(-1,-3)$.

- If the y-coordinates of both points are the same, then the line is parallel to the x-axis, for example, points $A(-3,1)$ and $B(4,1)$.

> If two distinct points lie on a line parallel to one of the coordinate axes, then the **distance** between the points is the **absolute value of the difference** of the unequal coordinates.
>
> - If $A(x_1,y_1)$ and $B(x_2,y_2)$ lie on a **horizontal line** (so $y_1 = y_2$), then $\textbf{distance(A,B)} = |x_1 - x_2| = |x_2 - x_1|$.
>
> - If $A(x_1,y_1)$ and $B(x_2,y_2)$ lie on a **vertical line** (so $x_1 = x_2$), then $\textbf{distance(A,B)} = |y_1 - y_2| = |y_2 - y_1|$.

Example: Points $A(-1,2)$ and $B(4,2)$ lie on a horizontal line. The distance between the two points is the absolute value of the difference of the x-coordinates. Thus,

$$\textbf{distance(A,B)} = |-1 - 4| = |-5| = 5$$

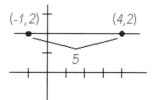

We get the same value when we subtract in the reverse order:

$$|4 - (-1)| = |5| = 5$$

Given two points on a line that is neither vertical nor horizontal, we make use of the Pythagorean Theorem:

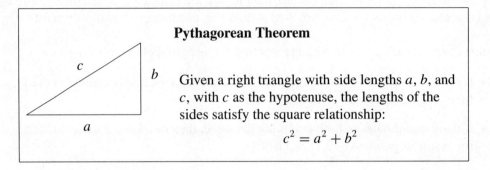

Pythagorean Theorem

Given a right triangle with side lengths a, b, and c, with c as the hypotenuse, the lengths of the sides satisfy the square relationship:
$$c^2 = a^2 + b^2$$

In order to find the distance between two points $A(x_1, y_1)$ and $B(x_2, y_2)$, we imagine the points as vertices of a right triangle. Then $a = |x_1 - x_2|$ and $b = |y_1 - y_2|$, so

$$\text{distance}(A,B) = \sqrt{(x_2 - x_1)^2 + (y_2 - y_1)^2}$$

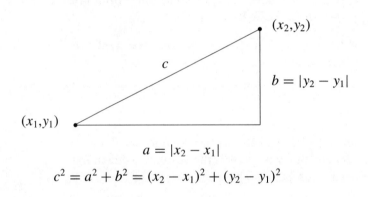

$$c^2 = a^2 + b^2 = (x_2 - x_1)^2 + (y_2 - y_1)^2$$

Distance Formula

The distance, d, between $A(x_1, y_1)$ and $B(x_2, y_2)$ is
$$d = \sqrt{(x_2 - x_1)^2 + (y_2 - y_1)^2}$$

1.2. DISTANCE AND MIDPOINT

Example: Given points A$(-2,1)$ and B$(3,2)$, the distance between A and B is given by

$$d = \sqrt{(3-(-2))^2 + (2-1)^2} = \sqrt{(5)^2 + (1)^2} = \sqrt{26}$$

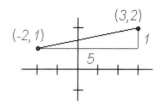

Note that when we apply the distance formula to two points that lie on a horizontal or vertical line, the formula reduces to the special case we considered at the start of this section.

A useful fact is that the Pythagorean Theorem works in reverse also: if the sides of a triangle satisfy the relation $a^2 + b^2 = c^2$, then we know that the triangle must be a right triangle! Thus, in order to verify that a particular triangle is a right triangle, we can find the lengths of its sides and check to see if they satisfy the Pythagorean relationship.

Midpoint

Now suppose we want to find the **midpoint** M(x,y) between two points A(x_1,y_1) and B(x_2,y_2).

> To determine the midpoint between two points (x_1,y_1) and (x_2,y_2), find the **average of the x-coordinates** and the **average of the y-coordinates** of the two points, giving
>
> $$M(x, y) = \left(\frac{x_1 + x_2}{2}, \frac{y_1 + y_2}{2}\right)$$

Example: Given the points A$(-2,1)$ and B$(3,2)$, the midpoint of the line segment connecting A and B is as shown.

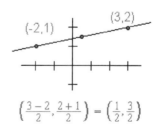

The applet on the web page demonstrates the distance formula and calculates the midpoint for two points that you select. Here you have some work to do before asking the computer to give you the result. Once you have selected two points, you calculate the distance between the points and the coordinates of the midpoint. Once you have a result, check it by clicking on Go *in the applet.*

Click on two points on the coordinate plane. The exercise will snap to the nearest points with integer coordinates. Find the distance between the two points and the midpoint of the line segment connecting them. Click Go to check your answers. If you want more problems, just click on Reset to start over.

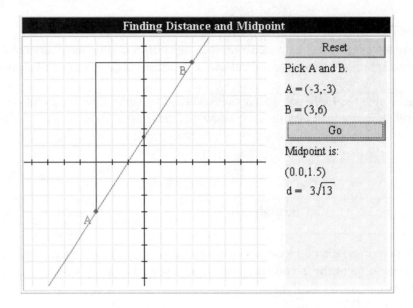

One of the applications of the distance formula is to find equations of circles; for exercises combining the distance formula and circles, see Section 1.4, Circles.

1.2. DISTANCE AND MIDPOINT

Additional Exercises

1. Determine the distance between each of the following pairs of points:

Points	Distance
(4,3), (4,7)	
(−11,7), (4,7)	
(−9,3), (4,7)	
(8,−2), (−3,5)	

2. Find the distance between each of the following pairs of points:

Points	Distance
(x,y), $(-x,y)$	
(x,y), $(-x,-y)$	
(x,y), $(x,-y)$	
$(-x,y)$, $(x,-y)$	

3. For what value(s) of x is the distance between $(x,3)$ and $(-2,7)$ equal to 5?

   $x = $ _____

4. Find the coordinates of the midpoint of the line joining $(-4,3)$ and $(0,-5)$.

   $(x,y) = $ _____

5. Find the midpoint of the line joining $(8,-3)$ and $(-3,5)$.

   $(x,y) = $ _____

6. The midpoint of the line joining $(x,-7)$ and $(4,y)$ is $(2,4)$. What are the values of x and y?

   $x = $ _____ $y = $ _____

7. Suppose you are packing your new 6 ft long skis into the trunk of your car. The trunk is 4.5 ft wide and 3 ft from front to back. Since the trunk is nearly full of suitcases, the skis must go in horizontally. Will they fit? What is the longest ski that could fit in the trunk?

1.3 Lines

A **linear equation** is an equation in which each term is a constant or a variable multiplied by a constant. Following are some examples of linear equations:

- $y = 2x + 5$
- $2x + 3y = 17$
- $-x + 8y = 3$

Linear equations in two variables are the simplest equations to graph; their graphs are straight lines. In this section we discuss how to make the transition from equation to graph and back. We will study properties of straight lines and their equations, such as x- and y-intercepts and slope. A given line may have more than one possible equation. Depending on the situation, one form may be more efficient than another, easier to find, or more informative. In this section we will also learn how to find and use different equations for a line.

> When we say that $2x + y = 4$ is an **equation of a line** we mean that *all* the points on this line have coordinates (x, y) that satisfy the equation $2x + y = 4$. The graph of this line is given on the right. What points can we find on this line? There are, of course, infinitely many. Some of them are the points $(-1, 6)$, $(1/2, 3)$, $(1, 2)$, $(3, -2)$. We can find points by substituting an x- or y-value into the equation and solving for the other variable. For example, setting $x = -2$ gives $y = 8$ so $(-2, 8)$ is on the line.

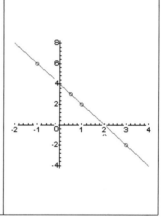

x- and y-intercepts

Any line that is not horizontal intersects the x-axis exactly once. The x-value at this point of intersection is called the ***x*-intercept**. The corresponding y-value must be 0 since the intersection point lies on the x-axis. We can find the x-intercept of a line from its equation by setting y equal to 0 and solving for x. Any line that is not vertical intersects the y-axis exactly once. The y-value at this point of intersection is called the **y-intercept**. The corresponding x-value must be 0 since the intersection point lies on the y-axis. We can find the y-intercept of a line from its equation by setting x equal to 0 and solving for y.

For example, consider the line given by $-x + 2y = 3$. To find the x-intercept, substitute $y = 0$ into the equation. This gives $-x = 3$, so $x = -3$ is the x-intercept.

Now go to the activity on the web page that asks you to determine the intercepts of a given line as well as to find two other points on the line. Make sure you read the instructions on how to enter the answers and follow them precisely!

The following activity shows some more equations of lines. To generate a new equation, just hit the [New] button. For each equation, find points on the line. In particular, find the x- and y-intercepts of the line, that is, the x- and y-values at which the line hits the axes, if these exist. Enter these in the form $x = c$ and $y = c$. Then enter the coordinates of two points on the line as pairs (x,y).

Finding Points on a Line

$-7x - y = -2$ [New] [Check]

- x-intercept
- y-intercept
- (x,y) on the line
- another such point

A crucial piece of information about any line is its **slope**. Given two points $A(x_1, y_1)$ and $B(x_2, y_2)$, we define the slope of the line containing A and B to be the value of the ratio

$$m = \frac{y_2 - y_1}{x_2 - x_1}$$

1.3. LINES

When you click on Slope *, a new window will appear as shown. In this demonstration, you'll click on two points in the plane. The coordinates of the points you have selected will appear. On paper, try to determine the slope of the line joining the two points. After you are done, press the* Go *button to see the value of the slope to compare it with yours. Use the* Reset *button to try another pair of points.*

To practice calculation of slopes, click here Slope .

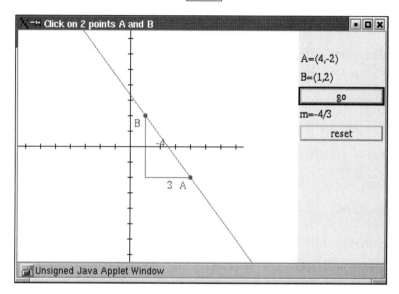

There are a number of things we would like to know about any straight line:

- An **equation** of the line
- The **slope** of the line
- The x- and y-**intercepts** of the line (if they exist)
- **Points** on the line that are not intercepts
- The equations of lines **parallel** to the given line
- The equations of lines **perpendicular** to the given line

Depending on the information we are given, we may use different methods to find an equation of the line. We say "an" equation, and not "the" equation, because the line may be given in terms of different equations. For example, the line described on page 11 is given via the equation $2x + y = 4$. It could also be described by $y = -2x + 4$, for example.

This last form, $y = -2x + 4$, is called the **slope-intercept** form of the equation because the slope and intercept can be seen immediately from the equation: in this case, the slope of the line is -2, and the y-intercept is 4. In general,

> Given an equation of the form $y = mx + b$, the slope of the line is m and the y-intercept is b.

A useful formula for finding equations of lines is the **point-slope formula**. If (x_0, y_0) is a given point on the line in question and (x, y) is any other point on that line, then the point-slope formula says $y - y_0 = m(x - x_0)$. This formula derives directly from the definition of slope given previously by replacing the points $A(x_1, y_1)$ and $B(x_2, y_2)$ with $A(x_0, y_0)$ and $B(x, y)$, respectively. For example, the line through the point $(3, -2)$ with slope 5 is given by $y + 2 = 5(x - 3)$.

Now go to the demonstration on the web page to see the effects of changing the slope and y-intercept on the steepness and placement of the line. Moving the left slider will change the slope; notice in particular the effect of changing the slope from positive to negative and back. Moving the right slider will change the y-intercept. What is the effect on the line?

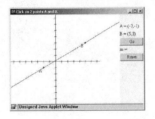

Move the sliders to see how changes in the slope m and y-intercept b affect the steepness and placement of the line. Also, as you vary the parameters, note the equation of the new line.

1.3. LINES

> In the special cases of a line parallel to one of the axes, the x-value or the y-value of all the points on the line is fixed. Thus, the line will have one of the following forms:
>
> - $x = a$ (parallel to the y-axis, or vertical)
> - $y = b$ (parallel to the x-axis, or horizontal)

Parallel Lines What is the crucial piece of information that ensures that two lines are parallel to one another? It's the fact that their slopes are equal to one another. How can we use this fact?

Suppose we are given a line $5x + y = 1$, and we are asked to find a line parallel to it. The first thing to realize is that there are *an infinite number* of lines parallel to a given line! So, we can choose any line that has the same slope as $5x + y = 1$. What is this slope? First translate the line to slope-intercept form: $y = -5x + 1$. Then take the coefficient of x: -5. This is the slope. So, any other line with slope -5 is parallel to the line $5x + y = 1$. For example, the line $y = -5x + 4$ satisfies the requirements.

Suppose now we are given the line $5x + y = 1$ and are asked to find the line parallel to it that passes through the point (2,3). Since we have discovered that the slope is -5, we are looking for an equation of the line with slope -5 that passes through (2,3). The equation will be of the form $y = -5x + b$, and we need to determine the value of b. Since we know that (2,3) is a point on the line, we may substitute the values $x = 2$ and $y = 3$ into the equation, getting $3 = -5(2) + b$. Solving for b, we get $b = 3 + 10 = 13$. Thus, the equation we are seeking is $y = -5x + 13$.

Perpendicular Lines What is the crucial fact that ensures that two lines are perpendicular to each other? It is the fact that the product of their slopes is -1. That is, if one line has slope 4 and another has slope $-1/4$, the product of the slopes is $4(-1/4) = -1$, so we know that the lines are perpendicular to one another. This is true no matter what the value of b is (in the slope-intercept version of the equation of the lines). How can we use this fact to find perpendicular lines?

Suppose we are given a line $2y + 4x = 1$. The first step is to find the slope of the line by translating its formula to the point-slope form. In this case, $2y = -4x + 1$, so $y = -2x + 1/2$. Thus the slope is -2, and any line whose slope is $-1/(-2) = 1/2$ will be perpendicular to our line. One such line would be $y = (1/2)x$. Another example is the line $x - 2y + 3 = 0$.

Now suppose we are asked for the line perpendicular to $2y + 4x = 1$ and with y-intercept 4. In this case we have no more to compute; we already know that a line perpendicular to $2y + 4x = 1$ will have slope $1/2$, and we can apply the y-intercept 4 to find the slope-intercept form of the line: $y = (1/2)x + 4$.

Summary of Methods for Finding Equations of Lines

1. **Given slope m and one point $A(x_1, y_1)$ on a line:**
   Since the value of m is given, we know we can write the line in the form $y = mx + b$. To find the value of b, we need only substitute the coordinates of the point A and the value of m into the point-slope form of the equation and solve for b.

   Alternatively, we may substitute the coordinates of the point A and the value of m into the point-slope form, and then convert to slope-intercept form if required.

2. **Given a point on a line and another line parallel to it:**
   Use the parallel line to find the slope of the required line, and then go back to method 1.

3. **Given a point on a line and another line perpendicular to it:**
   Use the slope of the perpendicular line to find the slope of the required line (the product of these slopes is -1), and then go back to method 1.

4. **Given two points on a line:**
   Use the two points to first find the slope of the line, and then go back to method 1.

5. **Given a point $A(x_1, y_1)$ on a line and the fact the line has undefined slope:**
   Since the line must be vertical, its equation is $x = x_1$.

You now have two options: You may go to the web page and click on the link examples in the next sentence to see examples of how we compute equations of lines given different types of information, or you can go to the web page and start practicing computing equations in the applet at the bottom of the page.

You can review examples of these types of problems or go right to the exercises below.

The final exercise on this web page requires you to find the equation of a line given two pieces of information. There are nine types of problems here, and you should practice each type several times. Put particular emphasis on the types of problems that you find more difficult!

In the exercise "Finding Equations of Lines" you will find a list of nine possible pieces of information about a line. To get your information about a line, either

1. Click on the $\boxed{\text{Any}}$ button, and the computer will choose the type of information at random; or

2. Click on the numbered button that corresponds to the type of information you'd like to practice. For example, if you'd like to calculate an equation of a line given two points on the line, click on button number 1.

1.3. LINES

Once you've computed an equation for the line, click in the white area to the right of the buttons. It will turn green, and you'll be able to type your answer. Make sure your answer is in the form of an equation!

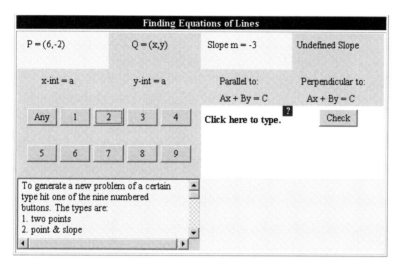

1.3. LINES

Additional Exercises

1. Which of the following are linear equations? Explain each "Yes" or "No."

    (a) $3x = 2y + 7$
    Yes   No   Reason: _____

    (b) $xy = 2$
    Yes   No   Reason: _____

    (c) $x^2 + y^2 = 1$
    Yes   No   Reason: _____

    (d) $4x + 7y - 3$
    Yes   No   Reason: _____

    (e) $2y + 11x = 7^2$
    Yes   No   Reason: _____

    (f) $y = (7x - 5)/3$
    Yes   No   Reason: _____

    (g) $y = (-2)^{-3}$
    Yes   No   Reason: _____

2. Which of the following points satisfy the equation $3x - 2y = 5$? Why or why not?

    (a) $(2,-1)$
    Yes   No   Reason: _____

    (b) $(1,-1)$
    Yes   No   Reason: _____

    (c) $(-3,8)$
    Yes   No   Reason: _____

    (d) $(-2,-5/2)$
    Yes   No   Reason: _____

3. Do the following points lie on the line $3x - 2y = 8$? Why or why not?

   (a) $(2,-1)$
   Yes , No   Reason: _____

   (b) $(1,-1)$
   Yes   No   Reason: _____

   (c) $(-4,-10)$
   Yes   No   Reason: _____

   (d) $(-2,-5/2)$
   Yes   No   Reason: _____

4. Find the x- and y-intercepts of the following lines.

   (a) $y - 4x = 2$
   x-intercept: (___, 0) y-intercept: (0, ___)

   (b) $x = 4$
   x-intercept: _____ y-intercept: _____

   (c) $y = -3$
   x-intercept: _____ y-intercept: _____

   (d) $x = 2y + 6$
   x-intercept: _____ y-intercept: _____

5. Determine the slope of each of the following lines.

   (a) $2x = 3y + 4$ Slope: _____

   (b) $7y + 2x = 0$ Slope: _____

   (c) $7y + 2x = -5$ Slope: _____

   (d) $6y = 11$ Slope: _____

1.3. LINES

6. Determine an equation for each of the following lines using the two pieces of given information.

Information	Equation
(a) Slope 3 and passes through $(-2,4)$	
(b) Slope -5 and passes through $(2,-1)$	
(c) Passes through $(4,1)$ and $(-5,2)$	
(d) Passes through $(-2,1)$ and $(-5,-7)$	
(e) Parallel to $3x = y + 4$ and passes through $(-3,4)$	
(f) Parallel to $y = 2x - 1$ and passes through $(3,-1)$	
(g) Perpendicular to $3x = y + 4$ and passes through $(3,1)$	
(h) Perpendicular to $y = -2x - 1$ and passes through $(-3,-4)$	

7. Suppose you are caught speeding in a 70 mph zone. The fine is $30 plus $10 per mph above the speed limit.

    (a) Write the dollar amount of the fine, F, in terms of speed, s.

    (b) If the fine is $370, how fast were you driving?

8. This time you are caught speeding in a 65 mph zone, and the fine is $40 plus $5 per mph above the speed limit.

    (a) Write the dollar amount of the fine, F, in terms of speed, s.

    (b) Use your formula to find the amount of your fine if you were driving 84 mph.

1.4 Circles

> A **circle** is the collection of all points in a plane that are a fixed distance from a given point. The fixed distance is called the **radius** of the circle, and the given point is the **center** of the circle.

For a circle drawn in a plane, if r denotes the radius and the center of the circle is at (h,k) then the equation of a circle in standard form (based on the distance formula) is
$$(x-h)^2 + (y-k)^2 = r^2$$

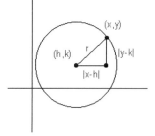

On the web page there is a demonstration showing the effects of changing the radius and location of the center of a circle. As you change the values of these quantities, the circle will be redrawn. The left slider controls the size of the radius, the bottom slider controls the x-coordinate of the center, and the right slider controls the y-coordinate of the center.

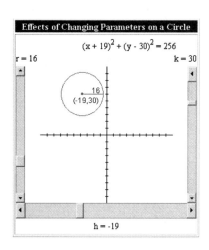

Move the sliders to see how changes in the radius r and center (h,k) affect the size and placement of the circle. Also, as you vary the parameters, note the equation of the new circle.

Now try the exercise on the web page that asks you to find the radius and center of a circle. For each question, you will be required to complete the square on the x terms and then separately on the y terms in order to put the equation in standard form.

In the following exercise, you are given a quadratic equation in x and y. First, complete the square in order to find the equation in standard form. Then use this form to find the center and the radius of the circle. When you are done with your calculations, click Graph it to check your answers and see the graph of the circle.

To begin, or to get a new equation, click on New .

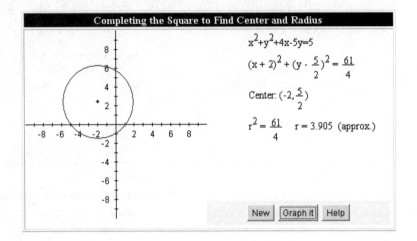

You now have an opportunity to apply your knowledge of circles to fixing a broken wheel. On the web, click on the Broken Wheel Problem link to go to the page with this applet.

A nice geometric problem that uses both circles and lines is sometimes called the Broken Wheel Problem and is presented in Section 1.6. Give it a try.

Sometimes we are not interested in the full circle but only in a **semicircle**. Let's find equations of four semicircles cut out of the circle on the right, which has the equation

$$(x-2)^2 + (y+1)^2 = 9$$

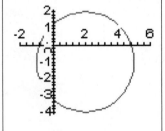

1.4. CIRCLES

To find the equation for the **upper semicircle**, solve the equation for $(y+1)^2$, then take the positive square root of both sides and solve for y itself.

For the **lower semicircle**, solve the equation for $(y+1)^2$, then take the negative square root of both sides and solve for y.

For the equation for the **left semicircle**, solve for $(x-2)^2$, then take the negative square root of both sides and solve for x.

For the **right semicircle**, solve for $(x-2)^2$, then take the positive square root of both sides and solve for x.

The following are the results of the computations of the upper, lower, right, and left semicircles in our example $(x-2)^2 + (y+1)^2 = 9$:

Upper Semicircle	Lower Semicircle
$y = -1 + \sqrt{9 - (x-2)^2}$	$y = -1 - \sqrt{9 - (x-2)^2}$

Right Semicircle	Left Semicircle
$x = 2 + \sqrt{9 - (y+1)^2}$	$x = 2 - \sqrt{9 - (y+1)^2}$

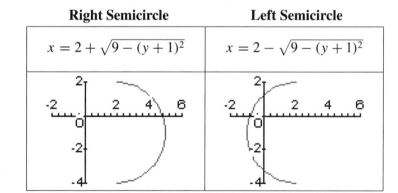

Now go to the web page to try the next exercise. For each question, you will be given the equation of a circle and then asked for the equation of a particular semicircle.

This next exercise requires you to find the formula for a given semicircle. Move the pointer over the ? in the typing area if you would like help in typing mathematics. Then click in the typing area (again) to begin typing.

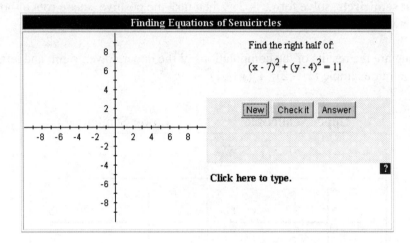

Additional Exercises

1. Determine the coordinates of the center and the radius of each of the following circles.

Circle	Center	Radius
$x^2 + y^2 = 4$		
$x^2 + y^2 + 6y + 7 = 0$		
$x^2 + y^2 - 4x + 2y - 4 = 0$		

2. Find all points on the circle $x^2 + y^2 = 9$ that have an x-coordinate of $1/2$.

    Points: _____

3. Find all points on the circle $(x - 1)^2 + (y + 2)^2 = 4$ that have a y-coordinate equal to -1.

    Points: _____

4. Find an equation for each of the following semicircles (notice whether you are asked for the upper or the lower semicircle):

Circle	Type	Equation
(a) $x^2 + y^2 = 9$	upper	
(b) $(x - 3)^2 + (y + 2)^2 = 1$	upper	
(c) $x^2 + y^2 - 4x + 2y - 4 = 0$	upper	
(d) $(x + 3)^2 + (y - 2)^2 = 1$	lower	
(e) $x^2 + y^2 - 2x + 4y - 4 = 0$	lower	

5. Find an equation of a circle of radius 2, centered on the x-axis and with a circumference that contains the point $(0,0)$.

    Equation: _____

6. Determine an equation of the circle that contains the points $(0,0)$, $(0,6)$, and $(3,3)$.

    Equation: _____

1.5 Complex Numbers

Complex Numbers

Complex numbers were developed as a result of the need to solve some types of quadratic equations. For example, the equation $x^2 + 1 = 0$ has no solutions in the real numbers. Trying to solve this equation would give $x^2 = -1$, meaning x would have to be a square root of -1. We define the **imaginary** number i to be a square root of -1 so that $i^2 = -1$. (The number i is sometimes denoted by the symbol $\sqrt{-1}$.) The other square root of -1 is written $-i$. Now the equation $x^2 + 1 = 0$ has the two solutions i and $-i$.

A **complex number** is a number of the form $a + bi$ where a and b are real numbers (note that one or both of them may be zero). The number a is called the **real part** of the complex number, and b the **imaginary part**. The number b is sometimes referred to as the coefficient of i.

If $b = 0$, then the complex number is just a real number. So, every real number may be thought of as a complex number. If a is zero, the numbers of the form bi are imaginary. Examples of complex numbers are $-3 + 2i$ and $0.3 - 0.257i$. Note that the coefficient of i in the last example is negative, and we wrote $0.3 - 0.257i$ rather than $0.3 + (-0.257)i$.

Operations with Complex Numbers

As with other sets of numbers, we can define arithmetic operations on complex numbers. The arithmetic of complex numbers is as follows: To add (or subtract) two complex numbers, we simply add (or subtract) the real parts and add (or subtract) the imaginary parts.

Example: Find the sum and difference of $-2 + 5i$ and $4 - 6i$.

Sum:
$(-2 + 5i) + (4 - 6i) = -2 + 4 + 5i - 6i = 2 - i$

Difference:
$(-2 + 5i) - (4 - 6i) = -2 - 4 + 5i + 6i = -6 + 11i$

The multiplication of two complex numbers is a little more complicated, but if you remember that $i^2 = -1$ and apply multiplication of binomials, it is not difficult.

Example: Find the product of $-2 + 5i$ and $4 - 6i$.

$$\begin{aligned}(-2 + 5i)(4 - 6i) &= -8 + 20i + 12i - 30i^2 \\ &= -8 + 20i + 12i + 30 \\ &= 22 + 32i\end{aligned}$$

Division of complex numbers is the most complicated of the four basic arithmetic operations. Assume we want to know what complex number is represented by $(c+di)/(a+bi)$, for $a+bi \neq 0$. That is, what complex number do we get when we divide $c+di$ by $a+bi$?

To describe this operation, we first need to define the **complex conjugate**.

> The **complex conjugate** of the complex number $a+bi$ is $a-bi$ (i.e., the opposite of the imaginary part).

The product of a complex number with its conjugate has a very simple form: $(a+bi)(a-bi) = a^2 - abi + abi - b^2 i^2 = a^2 + b^2$, which is a real number. We can use this fact to perform complex division by converting the denominator into a real number.

> To divide a complex number by another complex number, multiply the numerator and the denominator by the complex conjugate of the denominator.

Example: Divide $-2 + 5i$ by $4 - 3i$.

$$\frac{-2+5i}{4-3i} = \frac{(-2+5i)(4+3i)}{(4-3i)(4+3i)} = \frac{-23+14i}{25} = -\frac{23}{25} + \frac{14}{25}i$$

1.5. COMPLEX NUMBERS

Now go to the exercise set on the web page to practice arithmetic of complex numbers.

In this exercise set, you are given two complex numbers and are to perform the indicated arithmetic operation. Your answer should be a complex number. All answers involve integers or fractions in the real and imaginary parts of the answer. Do not convert to decimal notation.

The first four questions involve addition, subtraction, multiplication, and division, respectively. After that, the operation is random. Make sure that you try plenty of these problems.

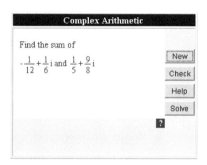

Additional Exercises

1. Determine the real and imaginary parts of each of the following complex numbers:

Complex Number	Real Part	Imaginary Part
$3 - 2i$		
$5i$		
7		
$\dfrac{-3 + 4i}{5}$		

2. Evaluate the following:

   (a) $(2 + 5i) + (-7 - 4i) = $ _____

   (b) $(-2 + 5i) - (-7 - 4i) = $ _____

   (c) $(-2 - 6i) - (7 + 11i) = $ _____

   (d) $(-1 + 8i) + (14 - i) = $ _____

3. Evaluate the following:

   (a) $(2 + 5i)(7 - 4i) = $ _____

   (b) $\dfrac{-2 + 5i}{-7 - 4i} = $ _____

   (c) $(-2 - 6i)(7 + 11i) = $ _____

   (d) $\dfrac{-1 + 8i}{3 - 4i} = $ _____

4. Find the value of k so that $2 + ki$ divided by $1 + 3i$ equals 2.

   $k = $ _____

5. If x is real, simplify $(x-i)(x+i)$ by multiplying out the product.

   $(x-i)(x+i) = $ _____

6. Let x and a be real numbers. Show that $(x-ai)(x+ai)$ is the same as $x^2 + a^2$.

1.6 Broken Wheel Problem

A famous problem that combines circles and lines is the **Broken Wheel Problem**. It appears in many forms: the wheel could be a gear, or a ceramic plate, or maybe a vase. The problem is usually stated something like this:

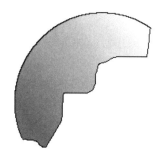

Suppose you are out digging around in the ground and you discover a fragment of an old wheel (insert here the circular object of your choice). Of course you wish to know the size of the intact wheel; that is, you want to find its original radius.

Here is one way to calculate the radius: Put the fragment down on a piece of graph paper and approximate the locations of three points on the circumference. Call those three points P_1, P_2, and P_3.

Now the fun begins!

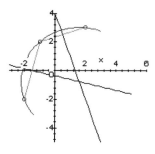

As an example, the circular arc in this figure represents the wheel fragment. The three points that have been located are $P_1 = (-2,-2)$, $P_2 = (-1,2)$, and $P_3 = (2,3)$. The line segments connect P_1 to P_2 and P_2 to P_3. Find the perpendicular bisectors of the line segments. Now find out where the bisectors intersect: that's the **center of the circle**. (In this case, it's at the point $(\frac{35}{22},-\frac{17}{22})$.) Now the radius can be found by calculating the distance from the center to any of the three original points. A complete solution to this example is given in Appendix B on page 331.

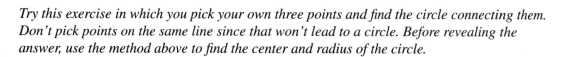

Try this exercise in which you pick your own three points and find the circle connecting them. Don't pick points on the same line since that won't lead to a circle. Before revealing the answer, use the method above to find the center and radius of the circle.

Try the example above or make up your own problems using the following Broken Wheel Fixer. Just click on three points you want on your circle and then hit the $\boxed{\text{Bisect}}$ button.

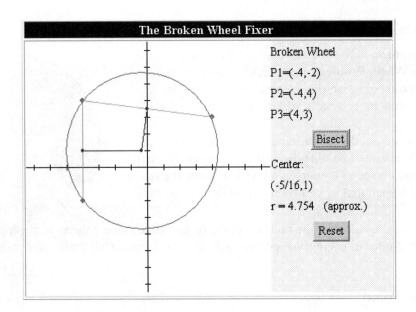

Module 2

Functions and Graphing

2.1 Functions

We often use the words "depends on" in everyday conversation, in sentences such as these:

- I don't know when I'll arrive; it depends on how bad the traffic is.
- The temperature of food depends on how long it has been in the oven.
- My course grade depends on my score on the final exam tomorrow.

In each case, we are saying that one value depends on another. In mathematics, we use the term "function" to represent the notion that the value of one quantity is dependent upon the value of another. Thus the following two statements express the same idea, although the first is admittedly the more conventional usage:

- My course grade depends on my score on the final exam tomorrow.
- My course grade is a function of my score on the final exam tomorrow.

In fact all three of those first sample sentences might be rephrased using the words "is a function of" instead of "depends on." Then all would have this structure:

The value of Y is a function of the value of X.

We can picture this dependence as a "function machine": A value of X is put into the machine, and the corresponding value of Y is output. What is hidden inside the machine is the rule that defines the correspondence between X and Y.

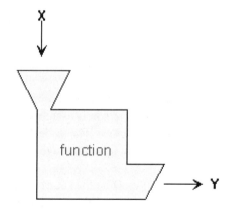

From the picture of the "function machine" we can see that important elements of a functions are as follows:

- What does X, the quantity we input into the machine, represent? It might be a time, a person, money, or just a number.
- What specific values of X are allowed? Perhaps some values of X should be omitted. For example, if X is a time measurement, maybe we should leave out negative values.

- What does Y, the quantity put out by the machine, represent?
- What specific values of Y can be output? For example, if Y represents a length or an area, it should never be negative.
- Just what is the rule hidden inside the machine?

In mathematics, the term "function" is defined so as to capture the notion of the dependence of one value or quantity upon another and to address the important aspects listed above.

> A **function** from the set D into the set C is a rule that assigns a unique element of C to each element of D. The set D is called the **domain** of the function, and C is called the **codomain**. The set of those elements of C that are assigned to elements of D is called the **range**.

Notice this definition of function does not use the word "number" since, in general, rules can be established between two sets even if they are not sets of numbers. In fact, if only sets of real numbers are involved, we use this wording for the definition:

> A **real-valued function** defined on a set of real numbers D is a rule that assigns a unique real number to each number in D.

These two examples illustrate numerical versus nonnumerical functions.

Example: Let D be the set of all persons living in your town. Let C be the set of all names of hair color. Each person is assigned a hair color. Since each person has a unique hair color, this establishes a function. The underlying rule is simply to get everyone's hair color so we don't have to make any up. The range contains those hair colors that actually appear in your town.

Example: Let D be the set of positive real numbers and C the set of real numbers. Each number x in D can be the side length of an x-by-x square. Assign to x the area of the square given by the formula x^2. Each side length produces a unique area, so this is a function. The range also consists of the positive real numbers since the square of a positive number is itself positive.

Whether a function is numerical or not, we think of the elements of the domain as "input" values and the corresponding range values as "output." This helps stress the idea that the rule defining the function takes a domain value, and from that, produces the corresponding range value. There are two ways we commonly write real-valued functions. The differences are purely notational since the same rule underlies each.

2.1. FUNCTIONS

We sometimes use one variable to denote the input and a second variable to denote the output, with a formula to show how the two quantities are related (if a formula is known). In the area example above, we used x to denote the side length. We could use A to represent the area of the square. In this simple rule, we know that $A = x^2$. We say that "A is a function of x." The input variable x is called the **independent variable**, and the output variable A is called the **dependent variable**. There is nothing special about the choice of letters for the variables, and we could even choose words for the variables (as is often done in writing computer code).

A second form of notation differs as follows: We assign a letter (such as f, g, etc.) to **name the function**. As before, we choose a letter to denote the input variable. If the function name is f and the input variable is x, then the **value of the function at x** (i.e., the output) is denoted by $f(x)$ (read "f of x"). If we know a formula relating the output to the input, we also write down that formula. So if we choose f to name the area function for the square, we write $f(x) = x^2$. Then, for example, $f(3) = 3^2 = 9$.

Whether we use just a variable name such as y for the output or the function notation $f(x)$ for the output, they would represent the same range value. In fact we often write $y = f(x)$, combining the two notations. One simple reason for combining the two is this: it provides the simplicity of the single variable name y for speaking and writing, plus the ability to use the $f(x)$ notation to instruct us to evaluate the function at a given value of x. Again using the area example: it is a lot easier to say or write "find $f(4)$" than to say or write "find the value of $y = x^2$ when x has the value 4." The next exercise asks you to do exactly this evaluation process for a variety of functions given by formulas.

Next on the web page is an exercise where you can evaluate a function at a given value. Function names and independent variable names differ in each question, but the process of evaluating the function is the same.

In this exercise, you are given a function and asked to find the value of the function at a given value. Please note the form for entering your answer: if you are asked for the value of $f(2)$ and this value is 8, then enter "$f(2) = 8$" as your answer.

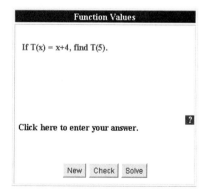

Before turning to graphing real-valued functions, a note on the word "unique" in the function definition: once a value is given for an independent variable x, there must be only one possible outcome $y = f(x)$—that is the "unique" part of the definition. For this reason, an equation of a circle does not represent y as a function of x. Consider the circle of radius 5, given by $x^2 + y^2 = 25$. Suppose, for example, that $x = 3$. Then $3^2 + y^2 = 25$, so $y^2 = 25 - 9 = 16$. At first glance, we might think that y should be 4. But it could also be -4, since $(-4)^2 = 16$. In other words, there is a value of x for which two different y's work. If we want y to be a function of x, we would have to select one of the two y-values to be the output for the input $x = 3$.

One such "problematic" value of x is sufficient to violate the requirements of a function, but in this circle example there are many more. Check it yourself by substituting values between 0 and 5 for x and finding different y's.

A **graph** of a function is essentially a drawing in the xy-plane that shows the inputs and corresponding outputs in the following sense: for each x in the domain of the function, with $y = f(x)$ its corresponding output in the range, mark the point (x, y) in the plane. Methods for obtaining the graph of a function are discussed in Section 2.2.

This is a good time to go to the web and to scroll through the following overview of functions. This will give you a glimpse of some of the types of graphs we will encounter in later sections. For now, just read through each step to become familiar with some of the terminology and basic ideas and shapes.

Now go through the next set of examples to get an overview of the concept of functions and their graphs, in particular the types of functions we'll be dealing with in this course.

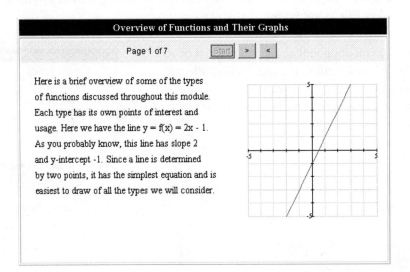

2.1. FUNCTIONS

There are two tests we can employ to help us decide whether a given equation describes a function:

- When you're given a graph of an equation, you may determine whether it describes a function by checking whether there is some vertical line that can be drawn that will intersect the graph at more than one point. This is called the **vertical line test.** If you find such a line, this means that you have found a value of x to which at least two values of y correspond—that is, there is no unique value of the "function." Therefore, the graph is not that of a function.

- **The algebraic equivalent of the vertical line test** is an algebraic determination for whether there is any value for x for which you can find two or more values for y. For example, consider the equation of a circle centered at the origin: $x^2 + y^2 = 8$. For the value $x = 2$, both $y = 2$ and $y = -2$ satisfy the equation. Thus this equation does not describe a function.

Go through this exercise to see if you understand exactly what the vertical line test says. In each step you will be given a graph of some sort (maybe not of a function!), and your task is to determine whether this is the graph of a function.

Which of these graphs of an equation describes a function? After you've answered correctly for one, click Next to get another question.

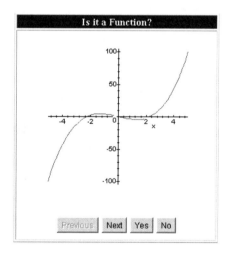

Notice that sometimes restricting the possible values of y or x or modifying the equation can give a function where there was none previously. For example, we may modify the equation of the circle of radius 3 centered at the origin, $x^2 + y^2 = 9$, to get $y = \sqrt{9 - x^2}$, which is a function whose graph is the upper half of that circle. The domain of a function, unless explicitly stated, is assumed to be the largest set of real numbers for which the function is defined. Determining the domain may not actually be an easy process. However, most of the functions we will deal with are given by a formula, and determining the domain is made easier by remembering the following rules:

- If the formula for the function has a denominator, then any value of the variable that makes the value of the denominator zero cannot be in the domain of the function.

- If the formula for the function contains a square root, then any value of the variable that makes the quantity under the square root a negative number cannot be in the domain of the function.

- If the formula for the function contains a square root in the denominator, then any value of the variable that makes the quantity under the square root a negative number or zero cannot be in the domain of the function.

Example: Review the following function:

$$f(x) = \frac{1}{x^2 + x}$$

This function has a denominator, so any values of x for which $x^2 + x = 0$ will not be in the domain of f. To solve this equation, we first factor to get $x(x + 1) = 0$. This equation has solutions 0 and -1. Hence the domain of f is all real numbers except 0 and -1.

Of course, if the formula for a function contains a combination of these types of terms, then you must determine the domain by considering each restriction and finding the common values of x. For example, if you find from one piece of the formula for the function that $x > 4$ and from another piece that $x > 2$, then the domain would be $x > 4$. As another example, suppose two parts of a formula tell you that $x > 0$ and $x < 5$, respectively. Then the domain of the function would be $0 < x < 5$.

For some other functions you will encounter in these modules, such as logarithmic functions (which we will study later), you will need to know the domain of the basic function as well as any restrictions listed above.

This quick check on domains of functions requires that you recognize that sometimes restrictions must be placed on the domain values to avoid problems such as division by zero or calculating a square root with a negative radicand.

2.1. FUNCTIONS

Here are some functions defined by formulas. Determine the domain of each function and click in the appropriate box. After you have the right domain for a function, click Next to go on to the next function.

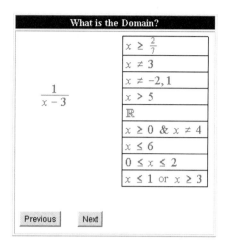

2.1. FUNCTIONS

Additional Exercises

1. Find $f(2)$ and $f(5)$ for $f(x) = 3x^2 - 7$.

   $f(2) =$ _____ $f(5) =$ _____

2. What is the domain of $f(x) = \sqrt{x-5}$?

   Domain: _____

3. What is the range of $f(x) = \sqrt{x-5}$?

   Range: _____

4. What is the domain of $f(x) = \dfrac{x}{x+4}$?

   Domain: _____

5. Explain why the circle described by $x^2 + y^2 = 16$ is not the graph of a function.

6. Two common scales for measuring temperature are the Fahrenheit scale (common in the United States) and the Celsius scale (common in Europe). Let F denote temperature measured in degrees Fahrenheit, and let C denote temperature in degrees Celsius. We will develop a conversion formula from Fahrenheit to Celsius as follows. (Water boils at 100°C, which is 212°F, and freezes at 0°C, which is 32°F.)

   (a) Find the equation of the line joining the points (100,212) and (0,32), and determine a formula that expresses temperature in degrees Fahrenheit as a function of temperature in degrees Celsius.

   (b) What temperature in Fahrenheit corresponds to 20°C?

   (c) At what temperature is the value in Fahrenheit the same as the value in Celsius?

7. An Internal Revenue Service booklet states that a single person having taxable income between $67,700 and $141,250 pays tax of $14,625 plus 30% of the amount over $67,700. Let T be the amount of tax such a person would pay, and let x denote his or her taxable income.

   (a) Determine a formula expressing T as a function of x.

   (b) How much tax will a single person pay if his or her taxable income is $103,000?

8. A window is in the shape of a square with a semicircle on top. Let A denote the area of the window, and let x be the width of the window. Express A as a function of x.

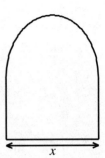

2.2 Graphing Techniques

What is the difference between the graphs of the functions $y = x^2$ and $y = x^2 + 4$? How about the differences between $y = x^2$ and $y = (x + 4)^2$? Instinctively it seems as though these graphs should be very similar, and indeed they are. However, in order to sketch the graphs we need a precise correlation between the algebraic differences and the graphical differences.

There are a number of concepts we will discuss in this section in order to learn how to sketch graphs: symmetry (even and odd functions), horizontal and vertical shifts, and vertical scaling (stretching and shrinking).

Symmetry

We'll begin with even and odd symmetry:

- A function $f(x)$ is called **even** if its value at x is the same as its value at $-x$; that is, if $f(x) = f(-x)$. An even function has a graph that is **symmetric about the y-axis**, which means you can take the graph for positive values of x and flip it over the y-axis to get the graph for negative values of x.

- A function $f(x)$ is called **odd** if its value at $-x$ is the negative of its value at x; that is, if $f(-x) = -f(x)$. An odd function has a graph that is **symmetric about the origin**, which means you can take the graph for positive values of x and flip it over the y-axis and then over the x-axis to get the graph for negative values of x. Equivalently, you can reflect this part of the graph about the origin, point by point, as is shown in the right-hand sketch below.

Check that the functions $y = f(x) = x^2$ and $y = g(x) = x^3 - 3x$ and their graphs demonstrate even and odd symmetries, respectively. That is, show that $f(-x) = f(x)$ and $g(-x) = -g(x)$ for all x. Their graphs are given below.

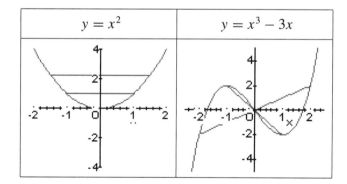

Four Example Graphs

These four functions are used in the examples and exercises that follow. Their graphs and basic properties should be memorized.

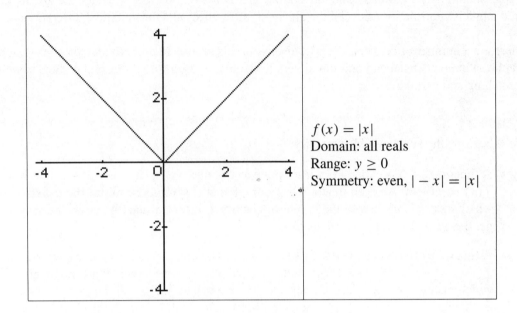

$f(x) = |x|$
Domain: all reals
Range: $y \geq 0$
Symmetry: even, $|-x| = |x|$

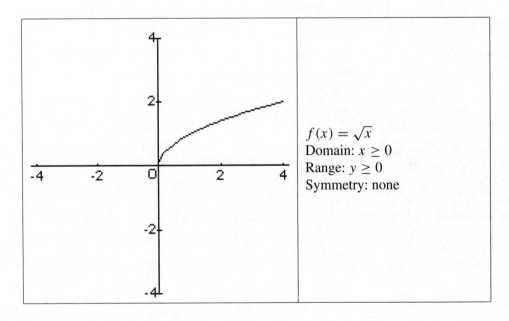

$f(x) = \sqrt{x}$
Domain: $x \geq 0$
Range: $y \geq 0$
Symmetry: none

2.2. GRAPHING TECHNIQUES

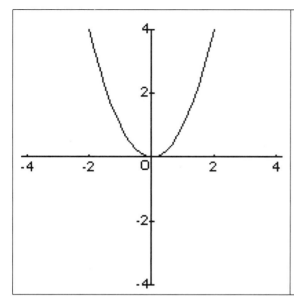

$f(x) = x^2$
Domain: all reals
Range: $y \geq 0$
Symmetry: even, $(-x)^2 = x^2$

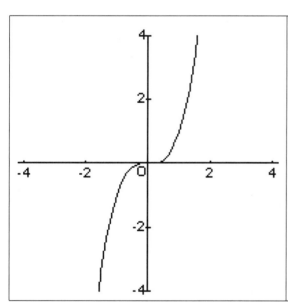

$f(x) = x^3$
Domain: all reals
Range: all reals
Symmetry: odd, $(-x)^3 = -x^3$

Shifting and Scaling

The three main transformations we will consider—horizontal shifts, vertical scalings, and vertical shifts—are described here with examples. Understanding these examples will help in the upcoming exercises.

Horizontal shift: The graph of $f(x - c)$ is the graph of $f(x)$ shifted horizontally $|c|$ units. The graph is shifted right if $c > 0$ and left if $c < 0$.

For $c = 2$, move right 2.

For $c = -3$, move left 3.

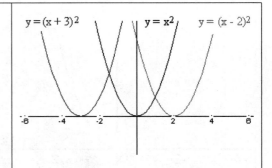

Vertical scaling: The graph of $kf(x)$ is the graph of $f(x)$ scaled vertically. If $|k| > 1$, the graph is stretched out from the x-axis and becomes steeper. If $|k| < 1$, the graph is shrunk or compressed toward the x-axis and becomes less steep. In addition, if $k < 0$, the graph is also flipped over the x-axis.

For $k = \frac{1}{3}$, less steep.

For $k = -2$, steeper and flipped over the x-axis.

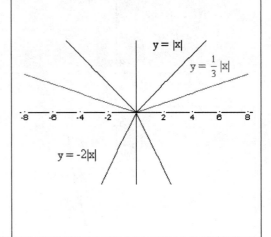

Vertical shift: The graph of $f(x) + c$ is the graph of $f(x)$ shifted vertically $|c|$ units. The graph is shifted up if $c > 0$ and down if $c < 0$.

For $c = 4$, move up 4.

For $c = -3$, move down 3.

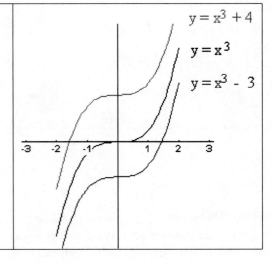

2.2. GRAPHING TECHNIQUES

Effect of Absolute Value

When a function is contained within **absolute value** bars, first consider the graph of the same function without the absolute value, and then flip the parts of the graph that are below the x-axis. Because, if $f(x) > 0$, then the absolute value bars have no effect on those points above the x-axis. If $f(x) < 0$, then the absolute value bars make that y-value positive, so those points below the x-axis are flipped above the x-axis. For example, here are the the graphs of $f(x) = x^2 - 3$ and $f(x) = |x^2 - 3|$:

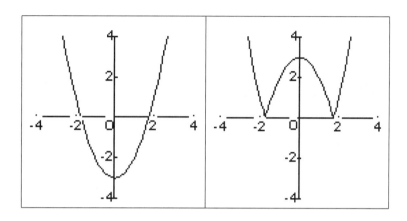

Examples and Exercises

When the actions described above are combined in one function, we produce the resulting graph by determining the cumulative effect of all actions. For example, the graph of $y = 2|x - 1| - 3$ can be drawn as follows:

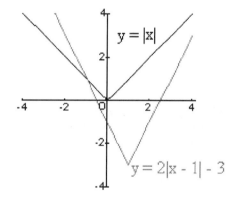

1. Start with the graph of $y = |x|$.

2. Move this graph right 1.

3. Make it steeper, since it has been scaled by 2.

4. Move the graph down 3.

By positioning the corner of the graph at the point $(1, -3)$ and noting that the y-intercept is -1 (that's another point on the graph we could plot), the correct steepness is displayed.

Go through a few examples in this activity, which demonstrates the effects of shifting and scaling on the basic shape of $|x|$. If you understand the connection between the manipulation of the function formula and the changes in the graph, you are ready to graph functions yourself in the next exercise.

In the following activity, you will see examples of graphs and the effects we have described above for the function $y = |x|$.

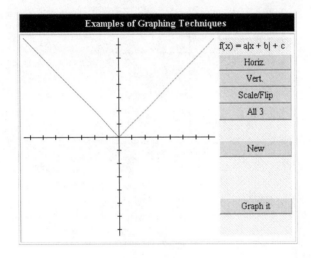

- Click on one of the effects: use Horiz. for a horizontal shift, Vert. for a vertical shift, Scale/Flip to scale the graph, or All 3 for a combination of these effects.
- Click New. You will see a new function using the effects you chose.
- Click on Graph it to see the graph of the new function.

You now have an opportunity to graph shifted and scaled functions on the computer. The transformations of the graph are done one at a time so that you can view the cumulative effects. Important: The order in which you do the transformations is crucial! Do several problems using each of the four function types. Also graph the function on paper to become efficient at this skill.

Now do the transformations yourself with a variety of functions and their graphs. Notice that you only need to know how to graph the original function in order to make the changes we have described here. The effects are cumulative so order is important. For example, a scaling followed by a vertical shift is not equivalent to a vertical shift followed by scaling.

2.2. GRAPHING TECHNIQUES

- Choose the type of function you want to graph by clicking on one of the four Function Type buttons.

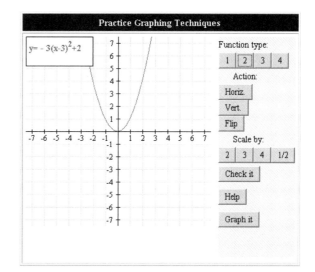

- To move the graph, click on Horiz. for horizontal shift or Vert. for vertical shift, then use the mouse to drag the graph.

- To flip the graph, click on Flip.

- To scale the graph, click on one of the four scaling factors to apply. Scaling is cumulative, so scaling by 2 and then by 3 scales the original function by 6.

- When you're done, click on Check it.

- To get a new function, click again on one of the four Function Type buttons.

- When you're done, you can click on Graph it to see the correct graph.

Additional Exercises

For each of the following questions, sketch the graphs of the given functions. If a question has more than one function, sketch all graphs on the same set of axes provided. You will need to choose appropriate scales for the sketches.

1. $y = |x|$, $y = \frac{1}{2}|x|$, and $y = -2|x|$

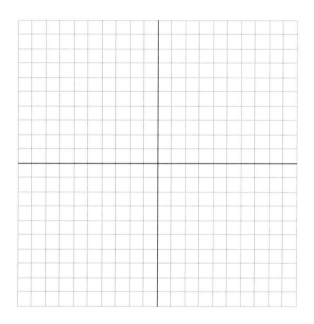

2. $y = \sqrt{x}$, $y = \sqrt{x-2}$, and $y = \sqrt{x+4}$

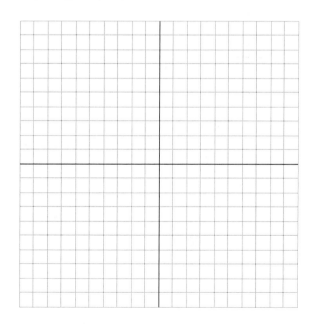

3. $y = x^2$, $y = x^2 + 1$, and $y = x^2 - 3$

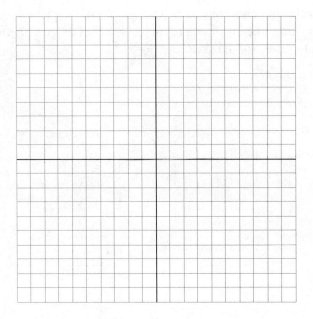

4. $y = 2(x - 3)^2 - 1$

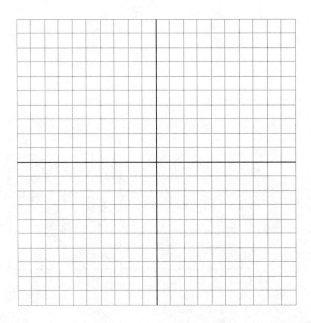

2.2. GRAPHING TECHNIQUES

5. $y = -3|x+1| + 2$

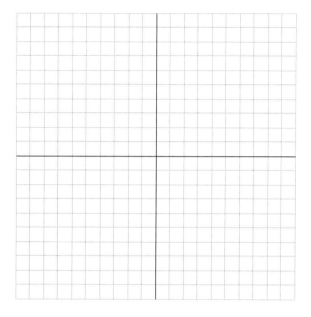

6. $y = (x+2)^3 - 3$

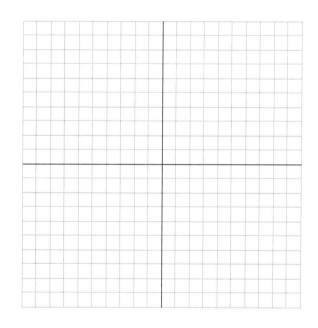

7. $y = -\sqrt{x-2} - 4$

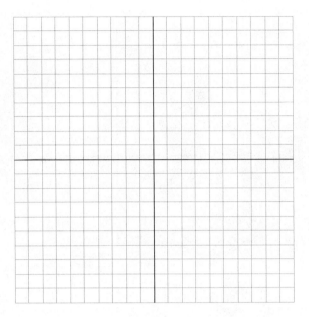

2.3 Quadratic Functions

So far, we have seen how to graph linear functions (i.e., functions whose graphs are straight lines). In this section, we will discuss a type of function that is, in a sense, the next step up. Straight lines are created by functions whose highest power of x is 1, and in this section we will discuss functions whose highest power of x is 2. These functions have graphs that are called **parabolas**.

> Functions of the form $y = ax^2 + bx + c$ (with non-zero coefficient a) are called **quadratic functions**, and their graphs are called **parabolas**.

Functions of this sort may be written in various ways, depending on our goal in each case. For example, consider the function $y = x^2 + 2x - 3$. It can be written in a number of ways; all of them represent the same function, and therefore all have the same graph.

- **General form:**
  $$y = x^2 + 2x - 3$$
- **Factored form:**
  $$y = (x - 1)(x + 3)$$
- **Standard form:**
  $$y = (x + 1)^2 - 4$$

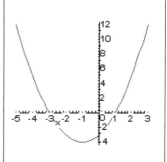

If you are not familiar with the technique of "completing the square" to get the standard form of a quadratic equation, you may use this opportunity to check it out. On the web, go to the link completing the square in the next paragraph. When you click on this link, the software will give you a pop-up window with instructions on how to carry out this technique as well as provide examples.

In order to get the general form of the quadratic function from either of the other forms, we need only multiply out and simplify, so this form is also called the **expanded form**. To get the factored form, we must first have the function in general form, and then factor the expression.

To get the standard form, we need to use the technique of completing the square; this too requires the general form as a starting point.

These are the four fundamental pieces of information that we need to sketch a parabola.

- The **direction** (whether the parabola opens up or down)
- The **vertex** of the parabola
- The **x-intercept(s)** (There may be zero, one, or two x-intercepts.)
- The **y-intercept** (There is always one.)

Direction

There are two basic shapes for the graph of a quadratic function. In most cases you will be able to deduce the **direction** of the parabola (i.e., whether it **opens up** or **opens down**) by applying the following rule for a quadratic in general form ($y = ax^2 + bx + c$):

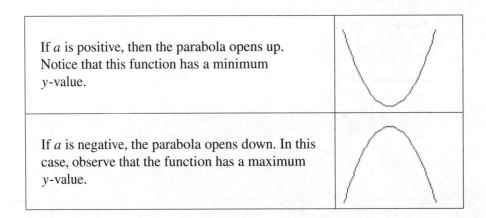

If a is positive, then the parabola opens up. Notice that this function has a minimum y-value.	
If a is negative, the parabola opens down. In this case, observe that the function has a maximum y-value.	

Vertex

A particularly important point in the graph of a quadratic is called the **vertex**. This point is either the maximum or the minimum point of the parabola. If the parabola opens down, the vertex is the maximum point, and if it opens up, then the vertex is the minimum. The coordinates of the vertex are most readily seen using the standard form of the quadratic function.

> The vertex of the quadratic $y = a(x - h)^2 + k$, is the point (h, k).

2.3. QUADRATIC FUNCTIONS

As an example, in the quadratic function we saw above, the standard form is $y = (x+1)^2 - 4$, so the vertex is at the point $(-1,-4)$.

Justification for the connection between the formula in standard form and the vertex comes from the graphing techniques we studied in Section 2.2. For the quadratic function $y = x^2$, the vertex is the origin, $(0,0)$. Subtracting h from x gives us a right horizontal shift by h units if h is positive, or a left horizontal shift by $|h|$ units if h is negative. Adding k to the rest of the expression gives us a vertical shift up by k units if k is positive, or a vertical shift down by $|k|$ units if k is negative. Thus the new vertex is at (h,k).

If we are given a quadratic function in general form, then to find the vertex we can either rewrite the expression in standard form or else use the following formula:

> If the quadratic function is given in general form, that is, $y = ax^2 + bx + c$, the x-value of the **vertex** is given by the formula $x = -b/2a$. The y-value of the vertex is found by substituting this into the formula for $f(x)$.

This formula is derived by rewriting $y = ax^2 + bx + c$ in standard form.

In the example above, $y = x^2 + 2x - 3$, this gives a vertex with $x = \frac{-2}{2} = -1$ and $y = (-1)^2 + 2(-1) - 3 = 1 - 2 - 3 = -4$. Thus, the vertex is again shown to be $(-1,-4)$.

x-intercepts

> An x**-intercept** is a point at which the graph of the function intercepts the x-axis. Algebraically, this means that $y = 0$.

The most convenient form of the quadratic function to use to find the x-intercepts is the factored form; we then set each of the factors equal to 0. In our example above, since the factored form of the function is $y = (x-1)(x+3)$, the x-intercepts are at $x = 1$ and $x = -3$, that is, at the points $(1,0)$ and $(-3,0)$. Notice that the x-value of the vertex is halfway between these, as we would expect.

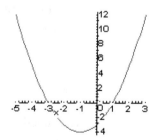

Not every quadratic function has two x-intercepts. There may be one or even no x-intercepts, as we will see:

$y = x^2 + 6x + 9$	When is there only one x-intercept? This happens when the factored form of the quadratic is a perfect square. For example, factor $y = x^2 + 6x + 9$.	
$y = x^2 + 1$	When are there no x-intercepts? This will happen when the quadratic function cannot be factored using real numbers: that is, when we would have to use complex numbers to factor it. For example, $y = x^2 + 1$ factors as $(x - i)(x + i)$.	

Another option for finding the x-intercepts is to use the **quadratic formula**. We use it when the function is given in general form: $y = ax^2 + bx + c$. In this case, the x-intercepts are given by

$$x = \frac{-b \pm \sqrt{b^2 - 4ac}}{2a}$$

y-intercept

> The **y-intercept** is the point at which the graph of the function intersects the y-axis.

Every quadratic has a (single) y-intercept. This is because the y-intercept is the point at which $x = 0$, and we can always substitute $x = 0$ into the quadratic function. Thus, the y-intercept of the quadratic function $y = ax^2 + bx + c$ is $(0, c)$. For the other forms of the function, just substitute $x = 0$ to find the corresponding value of y.

2.3. QUADRATIC FUNCTIONS

Additional Exercises

1. Write $y = x^2 - 8x + 27$ in standard form.

2. Write $y = x^2 - x - 4$ in standard form.

3. Find the vertex of $y = x^2 + 3x + 5$.

   vertex = (_____ , _____)

4. Find the x-intercepts of $y = 2x^2 - x - 7$.

   x-intercepts: _____

5. Find the y-intercept of $y = 3(x - 1)^2 - 8$.

   y-intercept: _____

6. Suppose a parabola has two real x-intercepts as given by the quadratic formula $x = \frac{-b \pm \sqrt{b^2 - 4ac}}{2a}$. Show that the value of x halfway between those intercepts is the vertex formula.

7. Newspapers often report the results of public opinion polls. A report might read something like "63 percent of our citizens are in favor of Proposition Z on the ballot." Usually included is a margin of error (MOE) that is often around 3 percent. The MOE comes from a formula that involves the true proportion x of the population that favor Proposition Z, which of course is unknown or there would be no need for a poll. A reasonable formulation for the MOE is $2\sqrt{\frac{x(1-x)}{n}}$, where n is the size of the sample polled and $0 \leq x \leq 1$ is the true proportion of the population who favor Proposition Z. One way to get around not knowing the MOE is to replace the actual MOE with an even larger number. In other words, if the MOE is really 2 percent, then reporting a MOE of 3 percent would not be reporting misleading results but rather playing it safe on the error estimate. How big could $2\sqrt{\frac{x(1-x)}{n}}$ be? Observe that the graph of the function $f(x) = x(1 - x)$ is a parabola that opens down. Its maximum value then occurs at the vertex. Verify that the vertex in fact occurs at $x = \frac{1}{2}$ and that $f(\frac{1}{2}) = \frac{1}{4}$. Replace $x(1 - x)$ with $\frac{1}{4}$ in the MOE formula and show that MOE $\leq \frac{1}{\sqrt{n}}$. You may have noticed that opinion polls often have a sample size of about 1000 individuals. Verify by calculator that $\frac{1}{\sqrt{1000}}$ is approximately 3 percent.

2.4 Maximizing the Area of a Rectangle

This demonstration presents a classic quadratic problem involving the construction of a rectangle of fixed perimeter that encloses a maximum possible area. Use the accompanying graph and drawn rectangle below to see the effect of changing dimensions upon enclosed area. A discussion of the set-up of the solution and the creation of a quadratic function follows the demonstration.

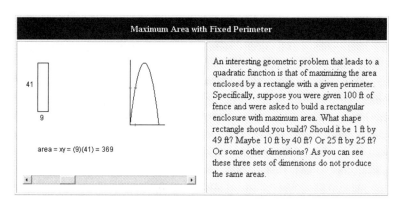

The quadratic function arises as follows: Let's use x and y as the dimensions of the rectangle, so the area is $A = xy$. To produce a function of the single variable x, we must use the constraint that perimeter $P = 2x + 2y = 100$. Here's how it's all put together:

- $A = xy$
- $2x + 2y = 100$, so $y = 50 - x$
- By replacing y, $A = xy = x(50 - x)$
- After expanding, $A = -x^2 + 50x$
- x should be between 0 and 50 to avoid negative A values.

Note that the resulting quadratic function has a negative leading coefficient, so this parabola opens down and indeed does have a highest point, which is the maximum possible value for A. Using the scroll bar in the illustration, we can change the x-value to any integer between 0 and 50. The corresponding rectangle is displayed, and its area is calculated. We also see a graph of $A(x) = -x^2 + 50x$, with the red dot locating the point $(x, A(x))$ on the graph. Since the vertex locates the highest point on this parabola, we can easily find the x-value by finding $-b/2a = -50/2(-1) = 25$. Aha! That's a square with area 625 ft^2. (Note that there is nothing special about the length 100 ft. We could start with any perimeter P and, following similar calculations, find that the largest area is attained when each side is $P/4$, forming a square.)

The next activity is a somewhat more complex version of the area problem. This time, the rectangle is further subdivided into sub-rectangles that alter the corresponding quadratic function. You can experiment with different designs and dimensions and try to detect the pattern before revealing the actual algebraic solution.

A variation of the fence problem is to require that, in addition to surrounding a rectangular area, the fencing must also be used to subdivide the area by running the fencing lengthwise or widthwise inside the rectangle. In this case, the total length of fence used is not the perimeter but the perimeter plus any lengths used to subdivide the area. In this next exercise, 600 ft of fence is available to construct and subdivide a rectangular region. Each time you choose a new value for the number of lengthwise sections built (M) and the number of widthwise sections built (N), a rectangle is displayed with the subdivision drawn in black. When $M = N = 2$, we have the previous exercise in which all fencing is used up in the perimeter. (The four buttons at the top of the applet change the position of the eye in case you need a different view to see the construction better.) Notice that as you change the shape of the rectangle by making it wider or narrower, the total length of fence used stays fixed at 600 ft as required. Experimentally find the dimensions that give the maximum area, and record those dimensions by pressing the Accept button each time you maximize the area. See if you can determine the formulas for the length and width that produce the largest area. Such formulas must of course involve M and N since the answer is, in general, not a square.

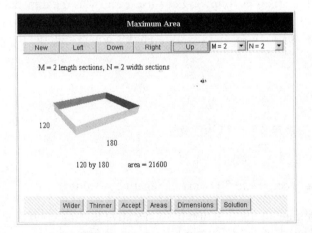

2.4. MAXIMIZING THE AREA OF A RECTANGLE

Additional Exercises

1. Find the dimensions of a rectangular fence enclosing maximum area using 800 ft of fence. What is the maximum area?

2. A rectangular area is fenced in and then subdivided once vertically and twice horizontally, producing six sub-rectangles. What dimensions will enclose the maximum area using 1000 m of fence? What is the maximum area?

3. A rectangular area is to be fenced. One side of the rectangle will be fenced with reinforced fencing that costs $12 per foot, whereas the other three sides will be made of fencing that costs $8 per foot. Find the maximum area that can be fenced for $2000 and the dimensions of the rectangle.

2.5 Graphing Quadratics

Which method we use to graph a parabola depends on the form of the equation we are given. For example, if the function is given in general form (i.e., $y = ax^2 + bx + c$), it is usually simplest to find the vertex of the parabola and its y-intercept and then to determine whether the parabola opens up or down. Once we know these facts, graphing the parabola is pretty straightforward.

For the parabola given by the quadratic function $y = ax^2 + bx + c$:

1. The x-value of the **vertex** is given by $x = -\dfrac{b}{2a}$.

2. The y-value of the vertex is obtained by substituting the x-value (from step 1) into the function.

3. The **y-intercept** is at c.

4. If $a > 0$, then the parabola **opens up**, and if $a < 0$, the parabola **opens down**.

To see an example, click the link in Section 2.5 on the web site.

Another option in graphing a parabola is to use the shifting and scaling techniques we saw in Graphing Techniques (Section 2.2). In order to use this process, the function must be in standard form: $y = a(x - h)^2 + k$. If it is not given in this form, you may use the technique of completing the square to bring the function to standard form (see Appendix A on page 329). Remember that the vertex is then at the point (h,k) and to scale the graph using a.

In the next activity you can practice graphing parabolas using both methods. To begin (or to move on to a new exercise once you've begun), click on the New button. The applet will guide you through the steps: entering the vertex, the y-intercept, and the x-intercept(s) (if there are any). At any stage of your work, you can click on Help to find out what to do next or on Solve if you want to see the full solution (for that stage).

Notice that you click on the appropriate points on the plane to enter the vertex and y-intercept, but you type in the x-intercepts in the workspace in the middle on the bottom. Don't forget to click in that space first to activate it!

Also be aware that you may get help with or solutions for either method by choosing the appropriate buttons: VF for vertex formula and CS for completing the square.

Practice graphing parabolas in this exercise using both methods: completing the square and using the vertex formula. Locate the vertex and all intercepts under either method. Sketch the graph on paper before revealing the graph on screen.

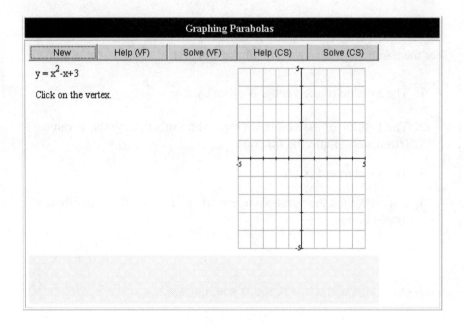

2.5. GRAPHING QUADRATICS

Additional Exercises

In questions 1 through 3, sketch the graphs of the given parabolas. Clearly label intercepts and vertices. You will need to choose appropriate scales for both axes.

1. $y = 2x^2 - 12x + 17$

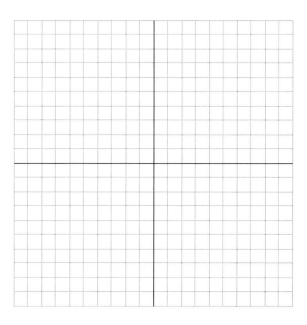

2. $y = -x^2 - 4x - 1$

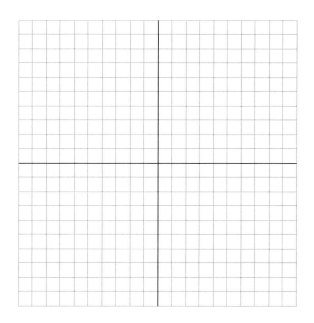

3. $y = 2x^2 - 3x - 5$

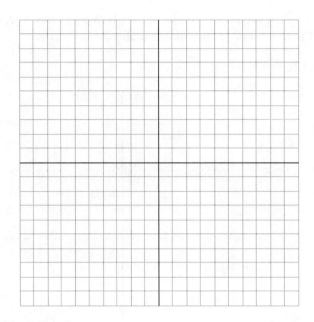

4. Find values for k so that the graph of $y = x^2 - 10x + k$ has

   (a) two x-intercepts

   (b) one x-intercept

   (c) no x-intercepts

Module 3

Operations and Inverses

3.1 Operations on Functions

Any two functions f and g can be combined in a variety of ways to produce new functions. Here we consider the arithmetic operations of addition, subtraction, multiplication, and division of functions and also the composition of functions.

Arithmetic Operations

Given the two functions $f(x) = x + 2$ and $g(x) = x - 2$, we can multiply to form the function $k(x) = (x + 2)(x - 2) = x^2 - 4$. We can also add them to form the function $h(x) = x + 2 + x - 2 = 2x$.

In general, we define the basic operations of addition, subtraction, multiplication, and division of functions as follows:

Operation	Function	Function Value
Sum	$f + g$	$(f + g)(x) = f(x) + g(x)$
Difference	$f - g$	$(f - g)(x) = f(x) - g(x)$
Product	fg	$(fg)(x) = f(x)g(x)$
Quotient	$\dfrac{f}{g}$	$\left(\dfrac{f}{g}\right)(x) = \dfrac{f(x)}{g(x)}$

Consider the functions $f(x) = 2x$ and $g(x) = 3x - 1$. Then,

- $(f + g)(x) = 2x + 3x - 1 = 5x - 1$
- $(f - g)(x) = 2x - (3x - 1) = -x + 1$
- $(fg)(x) = 2x(3x - 1) = 6x^2 - 2x$
- $\left(\dfrac{f}{g}\right)(x) = \dfrac{2x}{(3x-1)}$

We may use this concept in reverse to view a given function as some arithmetic combination of smaller ones. For example, the function $h(x) = x + 2/x - 2$ can be thought of as the result of dividing the function $f(x) = x + 2$ by the function $g(x) = x - 2$.

The domains of sums, differences, and products are all straightforward: these are defined wherever *both* of the original functions are defined, meaning that the domains of $f + g$, $f - g$, and fg will all be the intersection of the domains of f and g (i.e., those real numbers common to both domains). This is the set of values of x given by

$$D = \{x | f(x) \text{ and } g(x) \text{ are defined}\}$$

The domain of the quotient f/g cannot contain any values for which $g(x) = 0$. Thus the domain of f/g is the subset of D in which $g(x)$ is not zero. That is the set

$$D' = \{x | f(x) \text{ and } g(x) \text{ are defined}; g(x) \text{ is not zero}\}$$

Composition of Functions

Another way of combining two functions is by composition:

> For two functions f and g, we define the **composition of f and g** as $(f \circ g)(x) = f(g(x))$. This notation is read as "f circle g of x." $f \circ g$ is called the **composite** of f and g.

In other words, to find a formula for $(f \circ g)(x)$, take the formula for $g(x)$ and substitute this expression for x in $f(x)$. If x is assigned a numerical value, we can first find the number $g(x)$ and then plug that into $f(x)$. Or we can find a formula for $(f \circ g)(x)$ first, and then plug the value of x into that. It follows that the domain of $f \circ g$ is the set of all x in the domain of g so that $g(x)$ is in the domain of f.

3.1. OPERATIONS ON FUNCTIONS

Let's see some examples of composition:

If	Then
$f(x) = 2x - 1$, $g(x) = 4 - x$, and $x = -3$	$(f \circ g)(-3) = f(g(-3)) = f(7) = 13$, $(g \circ f)(-3) = g(f(-3)) = g(-7) = 11$, and $(f \circ f)(-3) = f(f(-3)) = f(-7) = -15$
$f(x) = x + 2$ and $g(x) = x - 2$	$(f \circ g)(x) = f(x - 2) = (x - 2) + 2 = x$, $(g \circ f)(x) = g(f(x)) = g(x + 2)(x + 2) - 2 = x$, and $(f \circ f)(x) = f(f(x)) = f(x + 2) =$ $(x + 2) + 2 = x + 4$
$f(x) = 7 - 2x$ and $g(x) = x^2 - 1$	$(f \circ g)(x) = f(x^2 - 1) = 7 - 2(x^2 - 1) = -2x^2 + 9$, $(g \circ f)(x) = g(7 - 2x) = (7 - 2x)^2 - 1 =$ $4x^2 - 28x + 48$, and $(f \circ f)(x) = f(7 - 2x) = 7 - 2(7 - 2x) =$ $-7 + 4x = 4x - 7$

In the following table are five pairs of functions. You are to obtain expressions for $(f \circ g)(x)$, $(g \circ f)(x)$, and $(f \circ f)(x)$ for each pair. Do this on paper. When you click in the appropriate box, the answer will be provided.

Calculate the entries in the following table. You can check your answers by clicking in the corresponding box.

Composition of Functions

$f(x)$	$g(x)$	$(f \circ g)(x)$	$(g \circ f)(x)$	$(f \circ f)(x)$
$\dfrac{1}{x}$	$3x - 2$			
$\sqrt{x}$	$3x - 2$			
x^2	$3x - 2$			
$5 - 2x$	$3x - 2$			
$\dfrac{1}{3}(x + 2)$	$3x - 2$			

Practice forming the composition of two functions in this next exercise. You are asked to form both $(f \circ g)(x)$ and $(g \circ f)(x)$.

Click on New in the applet below to get functions f and g. To enter the result of composing the functions, click in the area indicated to activate it. Remember that the order of composition matters and to enter your results in the correct space. If you need a reminder on how to type in mathematics, drag the mouse over the ? in the typing area, and a window with instructions will come up.

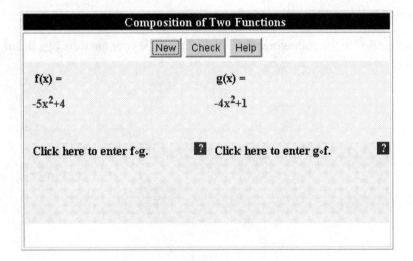

3.1. OPERATIONS ON FUNCTIONS

Since we often use variable names other than x and y and function names other than f and g, it is important to understand the composition process and notation with any names used. Here are some examples illustrating possibilities:

- Two functions can use different independent variables, for example, $f(x) = 2x + 1$ and $g(t) = t + 7$. Then $(f \circ g)(t) = 2(t + 7) + 1 = 2t + 15$. Notice that the independent variable in $f \circ g$ is t, which is the independent variable of the inner function g.

- One variable name may be an independent variable in one equation but a dependent variable in another equation. Suppose, for example, that the daily volume of water absorbed by a certain species of tree is given in liters by $w = h^2 + 3$, where h is the height of the tree in meters. In addition, the height is given by $h = 2t$, where t is the age of the tree in years. Then the composition $w \circ h$ produces $w = (2t)^2 + 3 = 4t^2 + 3$, which is water absorbed as a function of age.

Additional Exercises

1. If $f(x) = 3x^2$ and $g(x) = x^2 - 1$, then determine a formula for

    (a) $(f + g)(x) =$ _____

    (b) $(f - g)(x) =$ _____

    (c) $(g - f)(x) =$ _____

    (d) $(fg)(x) =$ _____

    (e) $\left(\frac{f}{g}\right)(x) =$ _____

    (f) $\left(\frac{g}{f}\right)(x) =$ _____

2. Determine the domains of each of the eight functions in the previous question.

3. If $f(x) = 7 - 1/x$ and $g(x) = 3$, determine $(f \circ g)(1), (f \circ g)(\frac{1}{7})$, and $(g \circ f)(2)$.

    (a) $(f \circ g)(1) =$ _____

    (b) $(f \circ g)(\frac{1}{7}) =$ _____

    (c) $(g \circ f)(2) =$ _____

4. If $f(x) = 7 - 1/x$ and $g(x) = 3$, determine $(f \circ g)(x)$ and $(g \circ f)(x)$.

    (a) $(f \circ g)(x) =$ _____

    (b) $(g \circ f)(x) =$ _____

5. Given $f(x) = x + 1$ and $g(x) = \sqrt{x - 1}$, determine $(f \circ g)(x)$ and $(g \circ f)(x)$.

    (a) $(f \circ g)(x) =$ _____

    (b) $(g \circ f)(x) =$ _____

6. Let $f(x) = -1/(x + 1)$. Find $f \circ f \circ f$. What is the domain of this function?

7. The volume of a sphere is $V(r) = \frac{4}{3}\pi r^3$ cm^3, where r is the radius in cm. If the radius is a function of time t and $r(t) = 4t$ cm after t minutes have elapsed, find the volume as a function of time by finding $V \circ r$.

8. The volume of a cube is $V(x) = x^3$ cm^3, where x is the side length in cm. If x is a function of time t and $x(t) = t^2 + 1$ cm after t seconds have elapsed, find the volume as a function of time by finding $V \circ x$.

3.2 Inverse Functions

A case of particular interest in the composition of two functions f and g is when $(g \circ f)(x) = x$ and $(f \circ g)(x) = x$. In this case we have the following definition:

> If f and g are two functions such that $(g \circ f)(x) = x$ for all x in the domain of f and $(f \circ g)(x) = x$ for all x in the domain of g, then we say that **g is the inverse of f**, and **f is the inverse of g**. We denote this by writing $g(x) = f^{-1}(x)$ and $f(x) = g^{-1}(x)$.

IMPORTANT: The -1 is not an exponent here! It is just part of the special symbol f^{-1}. Some examples you saw in the discussion of the composition of two functions are

- $f(x) = \dfrac{1}{x}$, where $(f \circ f)(x) = x$, so f is its own inverse.

- $f(x) = 3x - 2$ and $g(x) = \dfrac{x+2}{3}$, where $(g \circ f)(x) = x$ and $(f \circ g)(x) = x$ for all x, so the functions f and g are inverses of each other.

Note that if f and g are inverses of each other and we write $f(x) = y$, then $g(y) = x$. This emphasizes the relationship: if f sends the value x to the value y, then the inverse function must send that y right back to the x-value it came from. This is why we think of the inverse function as "undoing" whatever the function f did originally. We can see this effect in the second example above by noting that $f(2) = 4$ whereas $g(4) = 2$. Of course, we can exchange g and f in this discussion, that is, if we write $g(x) = y$, then $f(y) = x$.

Not every function has an inverse. The following characteristic of any function that has an inverse is critical:

> A function f with domain X and range Y is called **one-to-one** if different x-values in X correspond to different y-values in Y. This is equivalent to saying **if $f(x_1) = f(x_2)$, then $x_1 = x_2$.**

Compare the following two examples to get an idea why the one-to-one property is important in the discussion of inverses:

The function $f(x) = x + 5$ with domain all reals is one-to-one.	The function $f(x) = x^2$ with domain all reals is not one-to-one.
It seems pretty obvious that if we plug in different x-values, we get different y-values. That's the one-to-one property. For instance, $f(1) = 6$, whereas $f(2) = 7$, and so on. It sounds plausible that we might be able to construct a function g so that $g(6) = 1$ and $g(7) = 2$, and so on. Such a g would be the inverse of f.	Since $f(-2) = f(2) = 4$, we have a single y-value associated with two different x-values. That's not one-to-one. And we are going to have trouble coming up with an inverse g. How could we decide what $g(4)$ should be? It can't be both -2 and 2 because then g wouldn't be a function. That's a problem!

Do not confuse the one-to-one property with the definition of a function. Recall that if f is a function with domain X and range Y, then each x in X is associated with only one $y = f(x)$ in Y. If f is to have an inverse function g, each y in Y must be associated with only one x in X. We summarize all this as follows:

> A function f with domain X and range Y has an inverse g exactly when f is one-to-one. Moreover, the domain and range of g must be, respectively, the range and domain of f.

In the next activity you will check whether certain graphs are graphs of one-to-one functions by applying the horizontal line test: checking whether a horizontal line can intersect the graph at more than one point. Note that not every horizontal line must intersect at more than one point, but if even one does, then the function is not one-to-one.

3.2. INVERSE FUNCTIONS

A useful test to decide from the graph whether a function is one-to-one is the **horizontal line test**, which is comparable to the vertical line test for functions. Recall that the vertical line test says that a graph is that of a function only if any vertical line drawn in the plane intersects the graph at no more than one point. The corresponding horizontal line test says that a function is one-to-one only if each horizontal line intersects the graph at no more than one point.

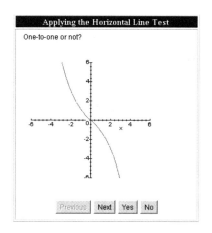

The next demonstration illustrates the reflection of points on the graph of a function to points on the graph of its inverse.

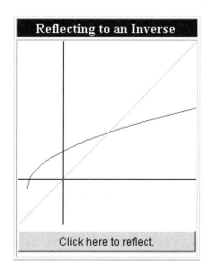

Suppose that f is a one-to-one function with inverse g and that (a,b) is a point on the graph of $y = f(x)$. Then we have $f(a) = b$. From the definition of an inverse function, $g(b) = a$, so the point (b,a) lies on the graph of $y = g(x)$. Thus to every point (a,b) on the graph of $y = f(x)$ there corresponds a point (b,a) on the graph of the inverse function $y = g(x)$. Consequently, the graph of $y = g(x)$ is the **reflection** of the graph of $y = f(x)$ **across the line $y = x$**.

In the following demonstration, you will enter a function and then see the graph of that function and its reflection. If the function is one-to-one, the reflection is of course the inverse graph. If the function is not one-to-one, notice that the reflected graph will fail the vertical line test.

In the following activity, you can choose a function and see both the graph of the function and the graph of its reflection on the same set of axes. Remember: If you choose a function that is not one-to-one, the reflection will not be a function. In the box at the top, enter the expression for the function you choose. In the x-vals box, enter the lower and upper limits of the domain you wish to use, separated by a space or a comma. Similarly, in the y-vals box, enter the lower and upper limits of the y-values to display. Click the Graph button to see the graph. With pencil and paper, sketch the reflection and then compare your graph with the one obtained when you click Inv.

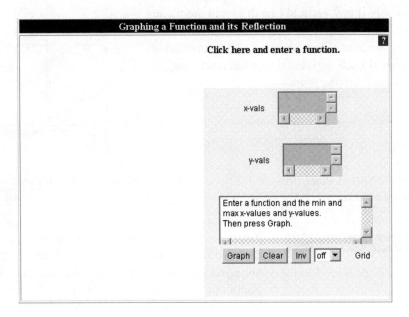

Knowing that a function has an inverse is one thing, but finding a formula for the inverse may be quite difficult; in fact, sometimes we may not be able to find a formula of the type $y = f^{-1}(x)$. The idea behind finding the inverse is, however, fairly simple:

> Since the graph of f^{-1} is obtained from the graph of f by switching the x- and y-coordinates of the points on the graph of f (that's reflection across the line $y = x$), the formula for f^{-1} is obtained by switching x and y in the equation $y = f(x)$ and then solving for y.

Of course, switching x and y is easy. However, we like to write the resulting equation in the form $y = f^{-1}(x)$, and that means solving for y. That's step 3 in the following example. This example is easily solved because the original formula was very simple. If f is more complicated algebraically, then step 3 can become more difficult or even impossible. Work through the steps in this example before proceeding to the exercises.

3.2. INVERSE FUNCTIONS

Step	Example: $f(x) = 5 - 2x$
1. Let y represent $f(x)$ by setting y equal to the expression for $f(x)$.	$y = 5 - 2x$
2. Let y represent $f^{-1}(x)$ by interchanging x and y.	$x = 5 - 2y$
3. Solve for y.	$y = \dfrac{5 - x}{2}$
4. y is now the formula for the inverse.	$f^{-1}(x) = \dfrac{5 - x}{2}$

Now find inverse formulas yourself by following the steps in the table above for each function given in the next exercise. Do all the required algebra on paper and enter the formula for the inverse function.

Now it's your turn. Follow the process above and find the inverse of the given function.

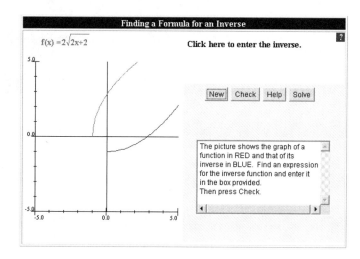

3.2. INVERSE FUNCTIONS

Additional Exercises

1. If $f(x) = x - 2$ and $g(x) = x + 2$, then determine a formula for $f \circ g$ and $g \circ f$. Are these inverse functions of one another? Explain.

2. If $f(x) = x^2$ and $g(x) = \sqrt{x}$, then determine a formula for $f \circ g$ and $g \circ f$. Are these inverse functions of one another? Explain.

3. Determine whether $f(x) = 4x - 5$ is a one-to-one function.

4. Determine whether $f(x) = x^3 - x$ is a one-to-one function.

5. Determine whether $f(x) = x^2 - x$ is a one-to-one function.

6. Find a formula for $f^{-1}(x)$ if $f(x) = 2x + 5$.

7. Find a formula for $f^{-1}(x)$ if $f(x) = \dfrac{3}{2x + 5}$.

8. Assuming $f(x) = \dfrac{x+1}{x}$ is one-to-one on some domain, find a formula for $f^{-1}(x)$. Determine the domain of f and the domain of f^{-1}. What is the range of f?

9. The graph of a one-to-one function is shown below. Sketch the graph of the inverse function.

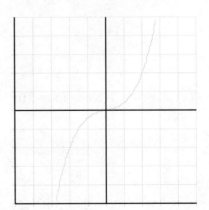

3.2. INVERSE FUNCTIONS

10. Find a formula for $f^{-1}(x)$ and graph both $f(x)$ and $f^{-1}(x)$. Give the domain and range of each.

    (a) $f(x) = \sqrt{x-2} - 1$

    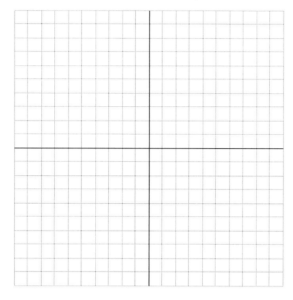

    (b) $f(x) = 2(x+2)^2 - 1$ for $x \geq -2$

    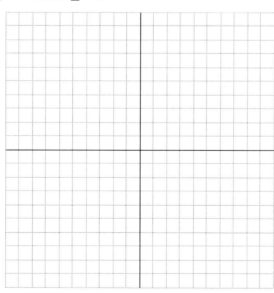

(c) $f(x) = 3\sqrt{x} + 2$

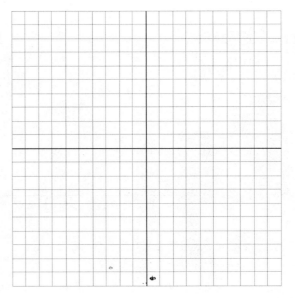

(d) $f(x) = (x-3)^2 - 2$ for $x \leq 3$

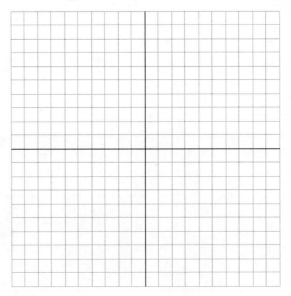

Module 4

Polynomial Functions

4.1 Polynomial Functions

So far we have discussed linear functions (e.g., $f(x) = 2x - 7$) and quadratic functions (e.g., $f(x) = -x^2 + 3x - 4$). In this section, we will extend our studies to functions that contain higher powers of x, for example,

$$f(x) = 2x(x-1)(x+1) = 2x^3 - 2x$$

As the powers of x increase, we find that the properties of the function become more complex. For example, consider the question of finding the x-intercept of a function. For a linear function, all we need to do is solve a simple equation for x. For a quadratic function, we must either use the quadratic formula or factor the polynomial. For higher-degree polynomials, we need additional techniques, and we will study these in this section.

First, let us see exactly what type of function we are dealing with:

> For some nonnegative integer n and numbers $a_n, \ldots, a_0$, a **polynomial** function has the following form:
>
> $$y = a_n x^n + a_{n-1} x^{n-1} + \cdots + a_1 x + a_0$$

> The coefficient a_n is called the **leading coefficient**, and a_0 is called the **constant term** of the polynomial. The number a_i is called the coefficient of x^i.

> Given a polynomial function such as the one above, other than the zero polynomial $y = 0$, the largest exponent n is called the **degree of the polynomial**.

Thus, the degree of a non-zero constant function, such as $y = 7$, is 0 (we consider 7 to be $7x^0$), whereas the degree of a linear function that is not constant, such as $y = 2x - 3$, is 1. The degree of a quadratic function is 2. In this section, we will discuss polynomial functions that have a degree greater than 2. (The constant function $y = 0$ is usually not assigned a degree.)

> If a polynomial is a product of two polynomials, then each polynomial in the product is called a **factor** of the polynomial. That is, if $f(x) = g(x)h(x)$, then g and h are factors of f.

> A **zero** of a polynomial, $p(x)$, is a value of x at which the polynomial takes on the value zero. Such a number is also a solution to the equation $p(x) = 0$ and is called a **root** of the equation.

Although the terms **zero** and **root** are intimately connected, we should take care to note the distinction between the two. The first is a value of x at which the corresponding y-value is 0, and the second is a solution of an equation. The following example should help clarify this.

Polynomial	$f(x) = x^3 + 2x^2 - x - 2$	Zeros	$-2, -1, 1$
Equation	$x^3 + 2x^2 - x - 2 = 0$	Roots	$-2, -1, 1$

Substitute the values given into the function to verify that these are indeed zeros!

Any discussion of polynomials must involve complex numbers. The simple example $y = x^2 + 1$ with zeros i and $-i$ shows that having real coefficients does not guarantee real zeros. Many of the important theorems in the upcoming sections are stated in terms of polynomials with complex coefficients and zeros. It helps to understand that, since the reals are contained in the same set of complex numbers as those numbers of the form $a + 0i$, polynomials with real coefficients do fall into the category of polynomials with complex coefficients. Since we are interested in graphing real-valued functions, we will mostly consider polynomials with real coefficients in the examples and exercises. This does not remove the possibility of complex zeros, however.

Both linear functions and quadratic functions are special cases of polynomial functions. For linear functions, $n = 1$, and we usually write $a_1 = m$ and $a_0 = b$. Thus, the general form of a linear function is usually written $y = mx + b$. For quadratic functions, $n = 2$, and we often write $a_2 = a$, $a_1 = b$, and $a_0 = c$. Thus, the general form of a quadratic function is often written $y = ax^2 + bx + c$.

4.1. POLYNOMIAL FUNCTIONS

This overview of polynomial functions illustrates some of the behavior of polynomial graphs. Scroll through the overview, reading each page and studying the accompanying graphs.

For a brief overview of polynomial functions, go through the steps in the following activity on the web:

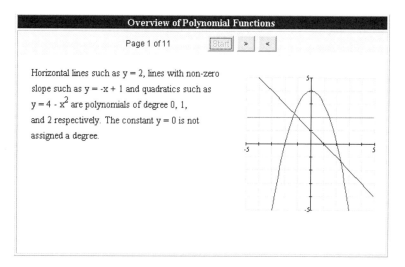

Graphing Polynomial Functions

The domain of a polynomial function is the set of all real numbers, and so the graph of a polynomial exists everywhere. In addition, to graph a polynomial function, we must find the following information, if possible:

- The x-intercept(s), if there are any
- The y-intercept
- Where the function is positive or negative
- The symmetries of the graph, if there are any
- The behavior of the function values for x large in magnitude

The steps in graphing a polynomial function are as follows.

Finding the x-intercepts: For a polynomial of degree n, there are between 0 and n x-intercepts. To find the intercepts, first factor the polynomial as far as possible, and use the factored form to find its zeros. You should be able to factor the function into linear or quadratic factors. This will tell you two things: what the zeros are and what the x-intercepts are. What is the difference?

Every x-intercept must correspond to a zero of the function, since that is a point at which the function has value 0. On the other hand, there may be zeros of the function that do not correspond to x-intercepts! Why not? Because they may be complex numbers, not real numbers. For example, the function $y = x^2 + 1$ has zeros i and $-i$, but neither of these is represented on the coordinate plane, so the function has no x-intercepts.

Section 4.3, Division, and Section 4.4, Zeros and Graphing, will be helpful in finding the zeros and x-intercepts of the function.

Finding the y-intercept: To find the y-intercept, substitute $x = 0$ into the function. Since this is always possible, every polynomial function has at least one y-intercept. Since the result of substituting (any) number into a function is unique, there is only one y-intercept.

Determining where the function is positive or negative: To do this, notice that the function can change sign only where it has an x-intercept. (The reason for this is given in the Intermediate Value Theorem, below.) Thus you need only determine the sign at one point in each interval between x-intercepts. In fact, the sign does not have to change at each x-intercept, so you must determine the sign in each of the intervals separately.

You may determine the sign by substituting a value of x in the interval you are checking and finding the sign of $f(x)$ there.

Determining symmetries: Sometimes the function you are graphing has symmetries that you can use to save work (or check your work) in graphing. For example, consider the function $y = x^3$. Suppose you know what it looks like for all positive x. Then notice that $f(-x) = -f(x)$ in order to sketch the graph for negative x. Once you have done all this, graph the function, while remembering the following points:

1. The graph of a polynomial function has no breaks or holes.

2. The graph of a polynomial function has no corners or sharp turns.

The following theorem is essential in the graphing of polynomial functions—it tells us, in essence, that there are no breaks in the sketch of such a function.

4.1. POLYNOMIAL FUNCTIONS

Intermediate Value Theorem for Polynomial Functions

If f is a polynomial function with real coefficients and $f(a)$ is not equal to $f(b)$, for some a less than b, then f takes on every value between $f(a)$ and $f(b)$ on the interval $[a,b]$.

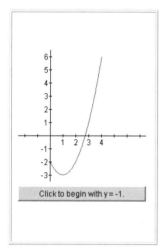

This is the graph of the function $f(x) = x^2 - 2x - 2$ on the interval $[a, b] = [0, 4]$. We have $f(0) = -2$ and $f(4) = 6$. The numbers -1, 3, and 5 all lie between -2 and 6 (along the y-axis). By the Intermediate Value Theorem there is at least one value of x, say $x = c$, in $[0, 4]$ for which $f(c) = -1$. There is another c for which $f(c) = 3$ and likewise for 5. Click on the demonstration to see where each value c is located. Notice that the c-value is found graphically by first drawing a horizontal line segment from the desired y-value on the y-axis to meet the graph over the interval $[0, 4]$. Then draw a vertical line segment from that point to the x-axis to locate c. So the Intermediate Value Theorem says that a horizontal line drawn at any y-value between $y = -2$ and $y = 6$ must hit the graph of $f(x) = x^2 - 2x - 2$ somewhere over the interval $[0, 4]$. It does not guarantee that we can actually solve for a value of c that corresponds to a selected y-value, simply that it does exist.

Determining the behavior for x large in magnitude: As values of x assume increasingly large magnitudes, the value of the polynomial function is controlled by the value of the largest power of x. For example, if $f(x) = x^5 + 7x^4 - x + 2$, then for an x that is very large in magnitude, the value of $f(x)$ is approximately the value of x^5.

For polynomials of even degree, this means that for an x that is very large in magnitude and negative, the function values will have the same sign as the function values for an x that is very large in magnitude and positive.

For polynomials of odd degree, for an x that is very large in magnitude and negative, the function values will have the opposite sign as the function values for an x that is very large in magnitude and positive. In particular, for odd-degree polynomials, the Intermediate Value Theorem tells us that at some number the graph crosses the x-axis. In other words, a polynomial of odd degree has at least one real zero.

4.1. POLYNOMIAL FUNCTIONS

Additional Exercises

1. Determine whether each of the following is a polynomial. If it is not, give a reason why. If it is, determine the degree, leading coefficient, and constant term.

   (a) $p(x) = -31$
   Degree ____ Leading coefficient ____ Constant term ____

   (b) $p(x) = x^3 + 3x^2 - 5x + 8$
   Degree ____ Leading coefficient ____ Constant term ____

   (c) $p(x) = x + 1$
   Degree ____ Leading coefficient ____ Constant term ____

   (d) $p(x) = 2^x - 3x + 2$
   Degree ____ Leading coefficient ____ Constant term ____

   (e) $p(x) = 4x^5 - 3x^7 + x$
   Degree ____ Leading coefficient ____ Constant term ____

   (f) $p(x) = x + \frac{3}{2}$
   Degree ____ Leading coefficient ____ Constant term ____

   (g) $p(x) = x^{-3} + x^2 - 2x$
   Degree ____ Leading coefficient ____ Constant term ____

   (h) $p(x) = x^{4/3} + x^2 - 2x$
   Degree ____ Leading coefficient ____ Constant term ____

   (i) $p(x) = 2(x - 3)^2$
   Degree ____ Leading coefficient ____ Constant term ____

2. For each of the following polynomials, find the value at the given values of x.

   (a) $p(x) = x^3 - 2x + 4$, $p(-2) = $ ____ $p(3) = $ ____

   (b) $p(x) = -x^3 - x^2 + 2$, $p(-1) = $ ____ $p(2) = $ ____

   (c) $p(x) = x^4 + x^3 - x^2 + 2x$, $p(-4) = $ ____ $p(0) = $ ____

3. An important use of polynomials is that of approximating other functions that cannot be evaluated exactly. For example, $f(x) = \sqrt{x+1}$ can be evaluated exactly only for certain x-values, like $x = 3$, for instance. Using methods of calculus, it can be shown that the polynomial $p(x) = \frac{1}{16}x^3 - \frac{1}{8}x^2 + \frac{1}{2}x + 1$ is close in value to $f(x)$ for x-values close to zero.

   (a) Verify that $p(0) = f(0)$. Thus for $x = 0$, the error in approximating $f(0)$ by $p(0)$ is zero. Another way to write this is $f(0) - p(0) = 0$.

   (b) Estimate the error for $x = 0.1$ this way. Use a calculator to find both $f(0.1)$ and $p(0.1)$ accurate to seven decimal places. Then find the difference $f(0.1) - p(0.1)$.

   (c) Estimate the error for $x = 0.2$ in a similar manner. Use a calculator to find both $f(0.2)$ and $p(0.2)$ accurate to seven decimal places. Then find the difference $f(0.2) - p(0.2)$.

   (d) Since $x = 3$ isn't nearly as close to zero as 0.1 or 0.2, we might expect less accuracy in our estimation. Without using a calculator, find both $f(3)$ and $p(3)$ exactly. Then find the difference $f(3) - p(3)$.

   (e) Graph both $f(x)$ and $p(x)$ together on a graphing calculator or using graphing software. Choose an interval that contains the number 0. For what x-values does $p(x)$ appear to be a very good approximation to $f(x)$?

4.2 Finding Zeros

Often we are given a polynomial but not its zeros. A number of methods can help us find the zeros of a polynomial. We list seven techniques and briefly describe how each helps.

Dividing Polynomials (Section 4.3)	A procedure to see whether one polynomial is a factor of another and, if it is, to see what its complementary factor is. In particular, to see whether $x - c$ is a factor of a polynomial.
Synthetic Division (Section 4.5)	A quick method to see whether $x - c$ is a factor of a polynomial.
Reducing the Degree of the Polynomial (Section 4.6)	Once you have a zero, you have a linear factor of the polynomial, which enables you to find additional zeros by working with a polynomial of lesser degree.
Descartes' Rule of Signs (Section 4.7)	Gives information on the possible number of real zeros that are positive in value and the possible number of real zeros that are negative in value.
Bounds on Real Zeros of a Polynomial (Section 4.8)	Methods for determining the largest positive value a zero of a polynomial can be. Also gives methods for determining the least negative value a zero of a polynomial can be.
Rational Zeros of a Polynomial (Section 4.9)	Gives a method for determining all possible rational zeros of a polynomial. This yields a list of values that you can check to see which are zeros.
Numerical Approximations to Zeros of a Polynomial (Section 4.11)	A method for finding the approximate value of a real zero.

4.3 Division

Division Algorithm for Polynomials

If $f(x)$ and $p(x)$ are polynomials with $p(x)$ not the zero polynomial, then there exist unique polynomials $q(x)$ and $r(x)$ such that

$$f(x) = p(x)q(x) + r(x)$$

In the above equation, the degree of $r(x)$ is less than the degree of $p(x)$, or $r(x)$ is the constant 0.

In the division of $f(x)$ by $p(x)$, the polynomial $q(x)$ is called the **quotient**, $r(x)$ is called the **remainder**, and $p(x)$ is called the **divisor**.

In the case that the divisor $p(x)$ is linear, that is, $p(x) = x - c$ for some number c, the degree of the remainder must be 0, or the remainder itself must be 0. In fact, we know even more:

Remainder Theorem

If a polynomial $f(x)$ is divided by $p(x) = x - c$, then the remainder is the number $f(c)$.

Now notice that if $f(c) = 0$, then $p(x)$ divides $f(x)$ exactly, and we have the following result:

Factor Theorem

A polynomial $f(x)$ has a factor $x - c$ if and only if $f(c) = 0$.

If we are given a polynomial with integer coefficients and a rational zero, a/b, it seems strange to see the factor written as $x - a/b$. However, we should realize that $bx - a$ is also a factor. For example, the polynomial $f(x) = 3x^2 + x - 2$ can be factored as $(3x - 2)(x + 1)$ or as $3(x - 2/3)(x + 1)$. Setting either $3x - 2 = 0$ or $x - 2/3 = 0$ and solving for x gives the same zero, $2/3$.

4.3. DIVISION

You should try this exercise now to make sure you understand the relationship between the zero value c and the factor $x - c$. For instance, $c = 2/3$ corresponds to $3x - 2$, whereas $c = -5/7$ corresponds to $7x + 5$. Practicing with this exercise now will help reduce the chance later on of making the common mistake of adding c instead of subtracting c from x to create the factor.

Try this quick quiz in which you are given zeros and asked to pick the corresponding factors.

Synthetic division is a technique to quickly find the quotient and remainder in dividing a polynomial by a linear polynomial. In order to learn how to do synthetic division, click on the link synthetic division.

In order to determine the quotient and remainder in the division of one polynomial by another, we may use long division or synthetic division (Section 4.5). We rely on synthetic division to divide by linear polynomials. By either method, the factor theorem tells us that finding factors is equivalent to finding zeros.

Additional Exercises

1. Use the Remainder Theorem to find the remainder if $f(x) = x^3 + 3x^2 - 5x + 8$ is divided by

    (a) $x - 3$ \qquad Remainder \underline{\qquad}

    (b) $x + 1$ \qquad Remainder \underline{\qquad}

    (c) $x + \frac{3}{2}$ \qquad Remainder \underline{\qquad}

2. Use the Remainder Theorem to find the remainder if $f(x) = x^3 + 3x - 1$ is divided by

    (a) $x - 3$ \qquad Remainder \underline{\qquad}

    (b) $x + 1$ \qquad Remainder \underline{\qquad}

    (c) $x + \frac{3}{2}$ \qquad Remainder \underline{\qquad}

3. Use the Remainder Theorem to find the remainder if $f(x) = x^3 + x^2 - 2x$ is divided by

    (a) $2x - 1$ \qquad Remainder \underline{\qquad}

    (b) $2x + 1$ \qquad Remainder \underline{\qquad}

    (c) $2x + 3$ \qquad Remainder \underline{\qquad}

4. Use the Factor Theorem to determine whether $f(x) = 6x^3 - x^2 + 7x - 12$ has a factor of

    (a) $x - 1$

    (b) $x + 2$

    (c) $x - \frac{1}{3}$

4.4 Zeros and Graphing

In our search for zeros of polynomials, we should have something that says that our efforts are not wasted seeking something that doesn't exist. The existence of zeros is guaranteed by this simple sounding yet deep and important theorem:

Fundamental Theorem of Algebra

Every polynomial of degree $n > 0$ has a zero.

When the Fundamental Theorem is combined with the Factor Theorem, we see that any polynomial $f(x)$ of degree greater than zero can be written as $(x - c)q(x)$ for some polynomial $q(x)$. But, if $q(x)$ is not constant, we may then apply the Fundamental Theorem to $q(x)$ to produce another factor of $f(x)$. Repeating this process until $q(x)$ reaches degree zero, we obtain the following factorization theorem:

Complete Factorization Theorem for Polynomials

Suppose $f(x)$ is a polynomial of degree $n > 0$. Then there exist n complex zeros of $f(x)$. That is, there are n (not necessarily distinct) complex numbers $c_1, c_2, \ldots, c_n$ such that

$$f(x) = a(x - c_1)(x - c_2)\ldots(x - c_n)$$

(a is the leading coefficient of $f(x)$).

It is not difficult to show that if a polynomial with real coefficients has a complex zero $a + bi$ (b not 0), then its conjugate $a - bi$ must also be a zero. Thus zeros of polynomials with real coefficients occur in conjugate pairs. If we want the factorization of such a polynomial to involve only **real numbers**, we can factor it into **linear** and **quadratic** factors, as we see in the following theorem:

Theorem on Factoring with Real Coefficients

Every polynomial with real coefficients of degree $n > 0$ can be expressed as a product of linear and quadratic polynomials with real coefficients in such a way that the quadratic factors cannot be further reduced over the reals.

The quadratic factors mentioned in the above result produce the conjugate pairs of any complex zeros. Note that each of the numbers $c_1, c_2, \ldots, c_n$ in the Complete Factorization Theorem is a zero of the polynomial. What if a particular number appears more than once? If, for example, a number appears three times in this list, then we name it once and say that it has

multiplicity 3. The following example illustrates the theorem. The polynomial $f(x) = 2x^4 - 4x^3 + 4x^2 - 4x + 2$ has degree 4 and leading coefficient 2.

Polynomial	$f(x) = 2x^4 - 4x^3 + 4x^2 - 4x + 2$		
Zeros	i	$-i$	1
Multiplicity	1	1	2
Zero Gives Factor	$x - i$	$x + i$	$(x - 1)^2$
Complete Factorization	$f(x) = 2(x + i)(x - i)(x - 1)^2$		

You can see from the table above that a polynomial can be reconstructed from its zeros and leading coefficient. You will do just that in the following exercise. You must use zeros to create factors of a polynomial and then enter the polynomial in factored form.

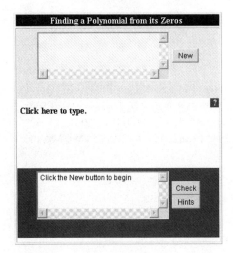

You now get to practice what was illustrated in the above example. For simplicity, only real number zeros will be considered in this exercise.

In this exercise, you will be given the complete list of zeros of a polynomial along with its leading coefficient, and you will be asked to write the polynomial in factored form. Note that you will be expected to spot multiplicities greater than 1.

4.4. ZEROS AND GRAPHING

We summarize the implication of multiplicity of a zero and how the lead coefficient, zeros, and their multiplicities relate to a polynomial.

> A polynomial of degree $n > 0$ has precisely n zeros when we count each zero according to its multiplicity. The factorization of a polynomial into linear factors, each with leading coefficient 1, is unique. Given the zeros, multiplicities, and leading coefficient of a polynomial, we can find the polynomial precisely.

We now turn our attention to the effect that multiplicity has on the graph of a polynomial. The multiplicities of a zero are important for the shape of the graph. For example, if a real zero has even multiplicity, the graph will touch the x-axis but not cross it; if a real zero has odd multiplicity, the graph will actually cross the x-axis (i.e., the polynomial will change sign).

Note that even though the complex, non-real zeros do not add x-intercepts, they too affect the shape of the graph.

Experiment with this pop-up demonstration to see the effects of multiplicities on the shape and behavior of a graph. In particular, note the connection between odd or even multiplicities and whether or not the graph crosses the x-axis.

Here is a demonstration to see the effect on the graph of a polynomial of adding zeros and multiplicities. Click the Open button to begin. In the pop-up window, click on the values you want to add as zeros; each time you click on a value, you are raising the multiplicity of that zero by 1.

Since you now know how zeros, x-intercepts, multiplicities, and degree can affect a polynomial graph, you should try the last exercise in this section. You must locate x-intercepts and the y-intercept, and decide whether the graph is above or below the x-axis before the graph of the given polynomial is revealed. Sketch the graph on paper before showing the graph on the computer.

Since knowing the factors of a polynomial gives the x-intercepts, it should be fairly easy to graph a polynomial when it is given in factored form. Give it a try in this exercise.

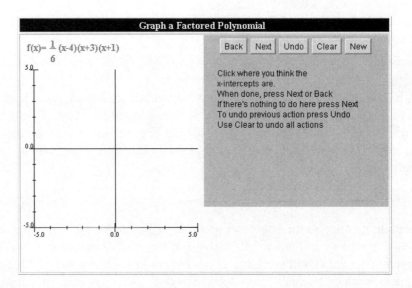

If you would like to practice graphing polynomials that are given in expanded form, go to Section 4.12, Factoring and Graphing.

Additional Exercises

1. Write the polynomial given its leading coefficient and zeros with multiplicities.

    (a) $a_n = -4$, zeros -7, $\frac{1}{5}$, and 9, with multiplicities 1, 3, and 2, respectively.

    (b) $a_n = 9$, zeros 0, $-\frac{1}{3}$, $\frac{1}{2}$, and 5, with multiplicities 2, 1, 1, and 4, respectively.

2. Sketch the graph of each polynomial.

    (a) $f(x) = (x+1)^2(x-1)^2(x-3)$

    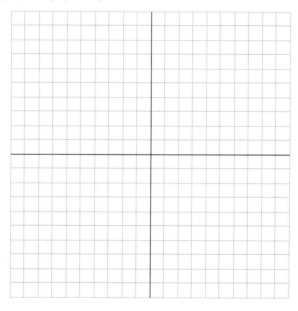

(b) $f(x) = -2x\left(x + \frac{3}{2}\right)\left(x - \frac{7}{4}\right)^2$

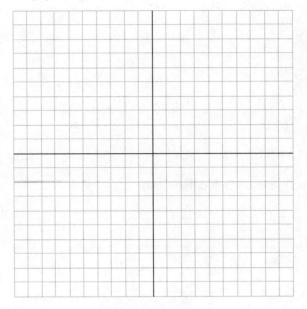

(c) $f(x) = \frac{1}{10}x^4(x+2)^2(x-3)^2$

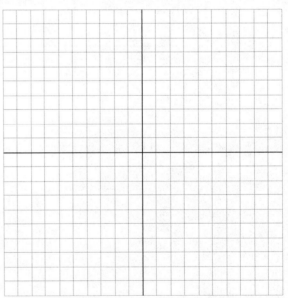

4.5 Synthetic Division

Synthetic division is an algorithm to find the remainder of dividing the polynomial

$$p(x) = a_n x^n + a_{n-1} x^{n-1} + \cdots + a_0$$

by the linear polynomial

$$d(x) = x - c$$

The linear polynomial that we are dividing by, $d(x)$, is called the **divisor**, and the polynomial we are dividing into, $p(x)$, is called the **dividend**. The Division Algorithm and the Remainder Theorem tell us that there is a polynomial $q(x)$, the **quotient**, and a real number r, the **remainder**, such that

$$p(x) = d(x)q(x) + r$$

The Synthetic Division Algorithm yields both the quotient and the remainder. The algorithm can be adapted to handle division by $bx - c$. However, the method *cannot* be used for nonlinear divisors. If the divisor has degree 2 or more, then we have to revert to the Long Division Algorithm.

Setting Up

Here are the steps involved in setting up the problem. As an example, suppose we want to find the remainder on dividing $x^5 + 4x^3 - 3x^2 + 3x - 1$ by $x + 2$.

We set up a table consisting of three rows. The number of columns required is two plus the degree of the dividend. In this case, since the dividend has degree 5, we will need a table with three rows and seven columns. In the first row, first column, we put the number that is being subtracted from x in $d(x)$. As $x + 2 = x - (-2)$, this entry is -2 and is shown in red on the web site in the table that follows. We will sometimes refer to this number as the **seed**. The remaining entries in the first row are the coefficients of the dividend written in descending order of powers of x. Any "missing" terms really have a coefficient of zero, and these must also be put in the row. These entries are in blue in the table on the web. Note the third column entry is 0 since the coefficient of x^4 in the dividend is 0 (a missing term).

-2	1	0	4	-3	3	-1
	0					
	1					

There are no other entries in the first column. The third row will contain the coefficients of the quotient and the remainder. We put a 0 in the second column of the second row. The second column of the third row is the sum of the two column entries above it, so we enter 1. Alternatively, it is the leading coefficient of the dividend divided by the leading coefficient of the divisor. This completes the table set-up.

It's best to see synthetic division in action to learn how the process works. You should step through this demonstration, reading the comments at each step, and make sure you understand the arithmetic being performed before proceeding. The steps you will see in the demonstration are described in Appendix C on page 333.

Algorithm

The process from now on is probably best described by example and involves a sequence of multiplications and additions. We have set up a table for a synthetic division problem. What are the dividend and divisor?

Now press the [Step] button to work through the algorithm.

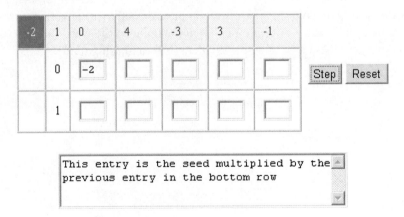

This entry is the seed multiplied by the previous entry in the bottom row

4.5. SYNTHETIC DIVISION

Finding the Quotient

In the following table, we have carried out the Synthetic Division Algorithm to find the remainder on dividing $x^3 + 1$ by $x + 1$. The third row entries, which appear on the web in green, tell us the coefficients of the quotient, whereas the last entry of the third row gives the remainder.

-1	1	0	0	1
	0	-1	1	-1
	1	-1	1	0

Since the quotient has a degree of one less than the dividend and the exponents are in descending order, the quotient is $x^2 - x + 1$. The remainder is 0.

The Case of $bx - c$

If the divisor is a linear polynomial, $bx - c$, where b is not zero, the synthetic division method can still be adapted to find the remainder. Clearly the case of $b = 1$ is what we have already discussed. Note that $bx - c = b(x - \frac{c}{b})$. Suppose that $q(x)$ is the quotient on dividing $p(x)$ by $bx - c$, and the remainder is r. Then,

$$p(x) = (bx - c)q(x) + r = b\left(x - \frac{c}{b}\right)q(x) + r$$

$$= \left(x - \frac{c}{b}\right)(bq(x)) + r = \left(x - \frac{c}{b}\right)q_1(x) + r$$

$q_1(x)$ is the quotient on dividing $p(x)$ by $x - c/b$. Hence the remainder is the same on dividing $p(x)$ by $x - \frac{c}{b}$ or $bx - c$; what changes are the coefficients of the quotient. To find the coefficients of the original quotient $q(x)$, we must divide the coefficients of the modified quotient, $q_1(x)$, by b.

As an example, let's work the problem where the dividend is $x^3 + x^2 + 1$ and the divisor is $4x - 1$. The completed table is shown next (make sure you get the same answers).

1/4	1	1	0	1
	0	1/4	5/16	5/64
	1	5/4	5/16	69/64

Thus when $x^3 + x^2 + 1$ is divided by $x - \frac{1}{4}$, the remainder is $\frac{69}{64}$ and the quotient is

$$x^2 + \frac{5}{4}x + \frac{5}{16}$$

So for $x^3 + x^2 + 1$ divided by $4x - 1$, the remainder is still $\frac{69}{64}$ but the quotient is

$$\frac{1}{4}x^2 + \frac{5}{16}x + \frac{5}{64}$$

Synthetic Division Calculator

The synthetic division calculator is one of the tools provided on the web. You may access this tool by clicking on the yellow button Open Tools *in the top bar. You will get a pop-up window with three buttons; the middle one is the button for the synthetic division calculator (the others are a calculator and a matrix manipulator).*

One of the tools available is the synthetic division calculator, which looks like this:

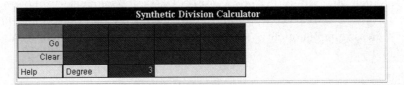

The calculator is used to determine the remainder and quotient upon dividing the polynomial $p(x)$ by the linear polynomial $x - c$. This calculator limits the degree of p to between 2 and 6, and it only allows the coefficients of p and the value of c to be integers or rational numbers written as the quotient of two integers. A decimal point is not allowed in the calculator. To use the calculator, first click on the third entry in the bottom row (it should originally be 3). The entry should change color. Enter the degree of the dividend here. If we use the example $x^5 + 4x^3 - 3x^2 + 3x - 1$, then you should type 5 in this rectangle. The number of columns in the calculator will change to accommodate the degree you entered. Now click on the leftmost

4.5. SYNTHETIC DIVISION

rectangle in the top row (it is colored red), and enter the seed value. Again, if we use, as an example, $x + 2$ as the divisor, you would enter -2 in this rectangle. Then either use the tab key or click with the mouse to enter the coefficients of the dividend along the top row. For our example, you would now enter 1, 0, 4, -3, 3, and -1. Once you have filled in all the entries, click on the Go button and the tool will calculate all the entries for you. The last entry in the third row is the remainder, and the other entries are the coefficients of the quotient. Remember that entries in the top row must all be integers or rational numbers written as the quotient of two integers (you cannot use a decimal point). For example, if the seed is one-half, then you enter it as 1/2. You cannot enter 0.5.

Practice performing the synthetic division process in this next exercise. You are required to enter the remainder in a division problem after using synthetic division to find the quotient and remainder. It is important to become efficient at this technique before proceeding to later sections.

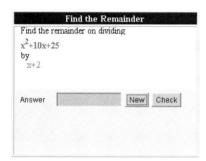

In this exercise you are given a polynomial and a linear divisor. You are to find the remainder on dividing the polynomial by the divisor. Your answer should be entered in the box in the lower half. Click on Check to check your answer and on New for a new problem. If you need help, use the synthetic division calculator above. Please note that to enter fractions in a text box, you must use the "/" key on your keyboard: for example, one-half is entered as 1/2.

4.5. SYNTHETIC DIVISION

Additional Exercises

1. Find the quotient and remainder when $f(x) = x^4 - 3x^3 - 4x^2 + x + 1$ is divided by

   (a) $p(x) = x - 2$ Quotient _____   Remainder _____

   (b) $p(x) = x + 1$ Quotient _____   Remainder _____

   (c) $p(x) = x - \dfrac{1}{2}$ Quotient _____   Remainder _____

   (d) $p(x) = 2x - 1$ Quotient _____   Remainder _____

   (e) $p(x) = 3x + 2$ Quotient _____   Remainder _____

   (f) $p(x) = 2x - 5$ Quotient _____   Remainder _____

2. Find the quotient and remainder when $f(x) = x^3 - 3x + 2$ is divided by

   (a) $p(x) = x - 2$ Quotient _____   Remainder _____

   (b) $p(x) = x + 1$ Quotient _____   Remainder _____

   (c) $p(x) = x - \dfrac{1}{2}$ Quotient _____   Remainder _____

   (d) $p(x) = 2x - 1$ Quotient _____   Remainder _____

   (e) $p(x) = 3x + 2$ Quotient _____   Remainder _____

   (f) $p(x) = 2x - 5$ Quotient _____   Remainder _____

4.6 Zeros and Factoring

Once we have found a zero of a polynomial, we can use it to simplify the factoring problem. This use is possible because once we have a zero of the polynomial, we actually know one of the linear factors of the polynomial and can divide the original polynomial by the linear factor to get a lower-degree polynomial.

Suppose we are asked to factor the polynomial $f(x) = 6x^2 + x - 2$, and suppose we have found out that $1/2$ is a zero of this polynomial. Then we know that the polynomial $p(x) = x - 1/2$ is a linear factor of $f(x)$. Now we divide $f(x)$ by $p(x)$. We know that the remainder will be zero, since $p(x)$ is a factor of $f(x)$. This division gives this result:

$$f(x) = p(x)(6x + 4)$$

Now, in order to find the zeros of $f(x)$, we need only find the zero of $6x + 4$. This is, of course, $-2/3$. Thus we now know that the zeros of $f(x)$ are $1/2$ and $-2/3$.

It is time to put the synthetic division process to work. Try this exercise in which you will find all zeros of a polynomial by factoring with the aid of synthetic division.

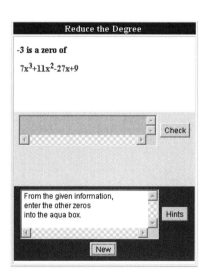

In the exercise, you will be given a quadratic or cubic polynomial and one of its zeros. Use this information to find the remaining zeros of the polynomial. Press New to begin, and enter your answer in the middle box.

Remember that if c is a given zero, then $x - c$ divides the given polynomial. In some cases, the given zero will be a fraction of the form a/b. For such problems you can use the fact that $bx - a$ is a factor of the given polynomial.

4.6. ZEROS AND FACTORING

Additional Exercises

1. Use the zero given to find the other zero of each of these quadratic polynomials.

    (a) $x = -3$ is a zero of $p(x) = 2x^2 + 5x - 3$.

    (b) $x = -\dfrac{3}{4}$ is a zero of $p(x) = 4x^2 + 11x + 6$.

    (c) $x = \dfrac{1}{5}$ is a zero of $p(x) = 15x^2 + 7x - 2$.

    (d) $x = -7$ is a zero of $p(x) = 3x^2 + 29x + 56$.

2. Use the zero given to find the other two zeros of each of these cubic polynomials.

   (a) $x = -2$ is a zero of $p(x) = 6x^3 - x^2 - 31x - 10$.

   (b) $x = 4$ is a zero of $p(x) = 15x^3 - 56x^2 - 51x + 140$.

   (c) $x = \dfrac{4}{7}$ is a zero of $p(x) = 21x^3 - 152x^2 + 31x + 28$.

   (d) $x = -\dfrac{5}{4}$ is a zero of $p(x) = 12x^3 + 79x^2 + 32x - 60$.

4.7 Descartes' Rule of Signs

Descartes' Rule of Signs

Suppose $f(x)$ is a polynomial with real coefficients and a non-zero constant term. Let N^+ be the number of sign changes in the coefficients of $f(x)$, and let N^- be the number of sign changes in $f(-x)$.

1. The number of positive real zeros of $f(x)$ is either
   (a) Equal to N^+ or
   (b) Less than N^+ by an even integer.
2. The number of negative real zeros of $f(x)$ is either
   (a) Equal to N^- or
   (b) Less than N^- by an even integer.

Descartes' Rule of Signs gives us an indication of the number of real zeros we might find for any given polynomial. Note that to find the possible number of negative zeros, we must look at changes of sign in $f(-x)$. A quick way to find $f(-x)$ is to change the sign of those terms in f having an exponent x that is an odd integer.

As an example, consider the following polynomial:
$$f(x) = x^4 - 2x^3 - 5x^2 + 9x - 4$$

Positive zeros:

$f(x)$ has three sign changes, so it has three or one positive zeros.

Negative zeros:

$f(-x) = x^4 + 2x^3 - 5x^2 - 9x - 4$ (or just change the sign on the terms with an odd exponent) has one sign change, so $f(x)$ has one negative zero.

Missing terms (i.e., zero coefficients) are ignored in applying Descartes' Rule of Signs. For example, $x^{50} - 1$ has one change of sign.

You will apply Descartes' Rule of Signs in this exercise to find possible numbers of positive and negative zeros of a polynomial. Check each box that indicates a number of zeros the given polynomial could have, not just the largest number possible.

In the next exercise you will be given a polynomial and asked to determine how many positive and how many negative zeros the polynomial can have. Check each number possible, not just the maximum number possible.

Practice Descartes' Rule of Signs

Use Descartes' rule of signs to determine how many positive and how many negative zeros the given polynomial could have. You are selecting how many zeros, not the values of the zeros themselves. You must select all possibilities. Use the check boxes to indicate your selections.

POLYNOMIAL

$2x^4+3x^3+4x^2-4x-5$

Pos. ☐ 5 ☐ 4 ☐ 3 ☐ 2 ☐ 1 ☐ 0

Neg. ☐ 5 ☐ 4 ☐ 3 ☐ 2 ☐ 1 ☐ 0

Check   New   Solve

Additional Exercises

Find the possible numbers of positive or negative zeros of each polynomial.

1. $f(x) = x^3 - 3x^2 - 4x + 1$

    Positive _____   Negative _____

2. $f(x) = -3x^3 + x^2 - 5x + 17$

    Positive _____   Negative _____

3. $f(x) = 2x^5 + 9x^4 - x^3 - 3x^2$

    Positive _____   Negative _____

4. $f(x) = x^4 - 2x^3 + 6x^2 - 8x + 11$

    Positive _____   Negative _____

5. $f(x) = 7x^4 - 3x^3 + 11x + 2$

    Positive _____   Negative _____

6. $f(x) = 3x^3 - 8x + 17$

    Positive _____   Negative _____

7. $f(x) = 2x^7 + x^4 + 7x^3 - 11$

    Positive _____   Negative _____

8. $f(x) = x^4 + 8x - 11$

    Positive _____   Negative _____

9. $f(x) = x^4 + 1$

    Positive _____   Negative _____

10. $f(x) = x^6 + 4x^2 - 15x - 17$

    Positive _____   Negative _____

4.8 Bounds for Zeros

One problem we'd like to answer is this: how big or small can the zeros of a polynomial be? That is, can we bound the values of all the real zeros of a polynomial? An initial bound that is easy and quick to calculate is the following:

> Given a polynomial $p(x)$, find the ratio of the largest magnitude of all the coefficients, $|a_i|$, to the magnitude of the lead coefficient, $|a_n|$, and add 1 to that result. If we call this number M, then any real zeros of $p(x)$ must lie between $-M$ and M.

The following examples show how this result on the bounds of zeros is used.

Polynomial $p(x)$	Largest $|a_i|$	$|a_n|$	M	Zeros Between
$f(x) = 2x^4 + 4x^3 - 7x^2 + 3x - 1$	7	2	$\frac{7}{2} + 1 = \frac{9}{2}$	$-\frac{9}{2}$ and $\frac{9}{2}$
$f(x) = 8x^4 + 4x^3 - 7x^2 + 3x - 1$	8	8	$\frac{8}{8} + 1 = 2$	-2 and 2

If synthetic division is used to find zeros, these upper and lower bounds can be refined as the factoring process progresses. The following theorem describes how this can be done.

> **Theorem on Upper and Lower Bounds on Zeros**
>
> Let $f(x)$ be a polynomial with real coefficients and a positive leading coefficient. Then,
>
> 1. Let u be a positive real number, and write
>
> $$f(x) = q(x)(x - u) + r$$
>
> where $r = f(u)$. If the coefficients of $q(x)$ are all nonnegative and r is positive, then the real zeros of $f(x)$ are not larger than u; that is, u is an upper bound for the real zeros of $f(x)$.
>
> 2. Let u be a negative real number, and write
>
> $$f(x) = q(x)(x - u) + r$$
>
> where $r = f(u)$. If the coefficients of $q(x)$, followed by r, alternate in sign (where 0 may be considered either positive or negative), then the real zeros of $f(x)$ are not smaller than u; that is, u is a lower bound for the real zeros of $f(x)$.

Before discussing this theorem in greater detail, we must realize that in both cases we are dealing with division by a linear polynomial. The coefficients of $q(x)$ and the value of r are, therefore, the values that appear in the third row of the table in the synthetic division process. Let's then restate the above theorem in terms of the third row of such a table.

> If u is a positive real number and the seed in a synthetic division problem, and if the third row of the table has only nonnegative values, then the zeros of $f(x)$ are no bigger than u.

> If u is a negative real number and the seed in a synthetic division problem, and if the values in the third row of the table alternate in sign, then the zeros of $f(x)$ are no smaller than u.

4.8. BOUNDS FOR ZEROS

As an example, we divide $f(x) = x^5 + 4x^3 - 3x^2 + 3x - 1$ by $x - 2$. We set up a table for synthetic division as usual.

The third row contains the coefficients of the quotient and the remainder. Since all of these entries are positive (and so nonnegative), 2 is an upper bound for this polynomial. That is, all the real number zeros of $f(x) = x^5 + 4x^3 - 3x^2 + 3x - 1$ are less than or equal to 2. In fact, they are strictly less than 2 since the remainder is not 0.

2	1	0	4	-3	3	-1
	0	2	4	16	26	58
	1	2	8	13	29	57

Knowing that 2 is an upper bound is not really enough; there are many upper bounds for the zeros (any number greater than 2 is an upper bound, for example). We are interested in finding the smallest possible upper bound, in order to narrow the field down as much as possible. Thus, we now ask whether 1 is an upper bound for the real zeros. How do we find out? We simply repeat the synthetic division using 1 as a seed instead of 2. We get the following:

1	1	0	4	-3	3	-1
	0	1	1	5	2	5
	1	1	5	2	5	4

Again the entries in the third row are all positive, so we know all real zeros of this polynomial are less than 1.

Now let's try to find a lower bound. This time we must use negative numbers as seeds in the synthetic division process. First, we try -1. Since the entries in the third row alternate in sign, -1 is a lower bound for the real zeros of our polynomial. We can conclude that if the polynomial has any real zeros, they are between -1 and 1.

-1	1	0	4	-3	3	-1
	0	-1	1	-5	8	-11
	1	-1	5	-8	11	-12

In this exercise you will use synthetic division and the technique described above to find upper and lower integer bounds on real zeros.

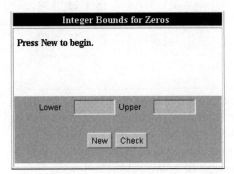

In these exercises, you will be given a polynomial. Use the theorem to find possible *integer* upper and lower bounds for the zeros of the polynomial. Recall that to apply the theorem, the lead coefficient must be positive. In those exercises where that is not the case, you should work with the negation of the given polynomial. Click on New to begin, and after using the theorem, enter your lower and upper integer bounds in the boxes provided. Remember, you are to find the *greatest* lower integer bound and the *least* upper integer bound. Then click Check to find how you did.

4.8. BOUNDS FOR ZEROS

Additional Exercises

1. Find an initial bound, M, for the zeros of each of the following polynomials using the polynomial coefficients.

   (a) $p(x) = x^{31} - 4x + 2$
   $M = $ _____

   (b) $p(x) = x^3 + 3x^2 - 5x + 8$
   $M = $ _____

   (c) $p(x) = x^2 - 3x + 2$
   $M = $ _____

   (d) $p(x) = 4x^5 - 3x^7 + x$
   $M = $ _____

2. What are the least integer upper bounds and greatest integer lower bounds predicted by the theorem on bounds for each of the following polynomials?

   (a) $p(x) = x^3 - 4x + 2$
   Least integer upper bound _____
   Greatest integer lower bound _____

   (b) $p(x) = x^4 + 4x^2 + x - 3$
   Least integer upper bound _____
   Greatest integer lower bound _____

   (c) $p(x) = -x^2 + 4x + 3$
   Least integer upper bound _____
   Greatest integer lower bound _____

3. Determine the least integer upper bound for the zeros of $p(x) = x^2 - 4x + 4$ using the theorem on refining upper bounds. Note that the largest zero of the polynomial is 2.

4.9 Rational Zeros

Theorem on Rational Zeros of a Polynomial

Suppose the following polynomial has integer coefficients:

$$f(x) = a_n x^n + a_{n-1} x^{n-1} + \cdots + a_1 x + a_0$$

also suppose that the rational number $\frac{c}{d}$ is a zero of $f(x)$, where c and d have no common factor other than 1. Then,

1. The numerator of the zero, c, is a factor of the leading constant term a_0, and

2. The denominator of the zero, d, is a factor of the leading coefficient a_n.

We can see that the only possible rational zeros of a polynomial must have the form

$$\frac{\textit{factors of the constant term}}{\textit{factors of the leading coefficient}}$$

We now illustrate this theorem and how it is used to find rational zeros of a polynomial.

Consider the polynomial $f(x) = 6x^2 + x - 2$.

The factors of the constant term -2 are ± 1 and ± 2.

The factors of the leading coefficient 6 are ± 1, ± 2, ± 3, and ± 6.

Candidates for rational zeros are ± 1, ± 2, $\pm \frac{1}{2}$, $\pm \frac{1}{3}$, $\pm \frac{1}{6}$, and $\pm \frac{2}{3}$.

Check which are zeros of the polynomial by using substitution or synthetic division. For example,

$$f(-1) = 6(-1)^2 + (-1) - 2 = 3$$

so -1 is not a zero of the polynomial. On the other hand,

$$f\left(\frac{1}{2}\right) = 6\left(\frac{1}{2}\right)^2 + \frac{1}{2} - 2 = \frac{3}{2} + \frac{1}{2} - 2 = 0$$

so $\frac{1}{2}$ is a zero of the polynomial.

You can practice determining whether or not particular rational numbers are possible zeros in this next exercise. It's a good idea to make sure you understand this procedure before continuing.

In the next activity you will practice finding candidates for zeros of polynomial functions. Click on ⟨New⟩ for a new function, and then click on each number that is a possible zero; that is, each number that satisfies the requirements of the theorem—that its numerator is a factor of the constant term, and its denominator is a factor of the leading coefficient (the coefficient of the highest power of x). The computer will assume that every number that you did *not* highlight is intended not to be a zero of the polynomial.

To check your answers, click on ⟨Check⟩. Your answers will be checked, and you will be told how many (out of how many possible) correct potential zeros you have picked, and how many "extras"—that is, how many numbers you have picked that cannot possibly be zeros of the polynomial given. If you want a hint as to what to do, click on ⟨Help⟩ in the activity.

If all your choices are correct, you may click on ⟨New⟩ to get a new polynomial. If not, you may click on a highlighted number to cancel it as a possible zero, or on a non-highlighted number to choose it as a possible zero. Do this as many times as necessary to get the correct combination.

Click on ⟨Answer⟩ to see the correct answers highlighted.

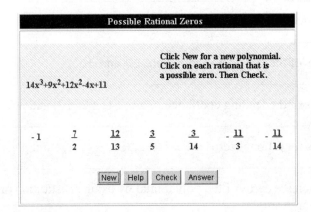

Now you can put all your skills together to find all the zeros of a given polynomial from scratch. You decide where to look, how many to seek, and what rational numbers to try. Use synthetic division to help locate zeros and factor the polynomial completely.

4.9. RATIONAL ZEROS

In the next exercise, you will be given a polynomial and asked to find all the zeros associated with it. Remember, these will be either integers or rational numbers.

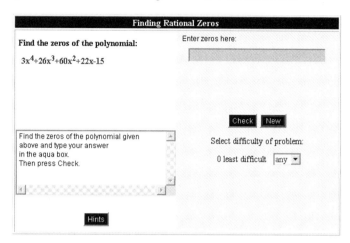

Additional Exercises

1. Use the Rational Zeros Theorem to list the possible rational zeros of each polynomial. Then use that list and synthetic division to find all zeros.

    (a) $p(x) = 4x^3 + 15x^2 - 31x - 30$

    (b) $p(x) = 15x^3 + 23x^2 - 18x - 8$

(c) $p(x) = 6x^4 - 71x^3 + 84x^2 + 275x - 150$

(d) $p(x) = 8x^4 - 46x^3 + 7x^2 + 121x - 30$

4.10 Complex Zeros and Coefficients

> **Theorem on Conjugate Complex Zeros of Polynomials with Real Coefficients**
>
> Suppose $f(x)$ is a polynomial with real coefficients and that it has a complex zero $c = a + bi$. Then the **conjugate** of c, $a - bi$, is also a zero of the polynomial.

Consequently, there must always be an even number of complex zeros in a polynomial with real coefficients. So, a polynomial with odd degree must have at least one real zero.

We illustrate this theorem with two examples.

Example: Consider the polynomial $f(x) = x^2 + 1$. Clearly, this has no real zeros, since the equation $x^2 = -1$ has no real solutions. However, the imaginary number i is a solution, hence also $-i$ is a zero of $f(x)$.

The second example is more involved and shows how we might go about completely factoring a polynomial.

Example: Find all the zeros of the polynomial $f(x) = x^3 - 4x^2 + 6x - 4$.

Solution: First, we notice that the degree of the polynomial is 3, which is an odd number. Hence there will be at least one real zero for this polynomial. The possible rational zeros here are $1, -1, 2, -2, 4$, or -4. After checking them, we find that 2 is a zero of this polynomial. Now that we have one zero of the polynomial, we divide $f(x)$ by the linear factor that comes from this zero, $x - 2$ (see Reducing the Degree of a Polynomial in Section 4.6), and find

$$f(x) = (x - 2)(x^2 - 2x + 2)$$

In order to find the rest of the zeros of $f(x)$, we need only find the roots of $x^2 - 2x + 2 = 0$. Using the quadratic formula, we find two conjugate complex roots $1 + i$ and $1 - i$. The complete factorization of $f(x)$ is thus

$$f(x) = (x - 2)(x - (1 + i))(x - (1 - i))$$

Note that once we have a quadratic factor for a polynomial, using the quadratic formula will always give us both complex zeros, since the conjugates are "built into" the formula.

The exercise provided on the web page requires you to find the zeros of a cubic polynomial. You may select the difficulty of the problems presented. Try several at level 0 (the easiest), and then try to build up to the highest level of difficulty.

In the following exercise, you are asked to find the zeros of a cubic polynomial. From the discussion above, we know that at least one zero is real. In the questions you are given, there will be exactly one real zero and two complex zeros.

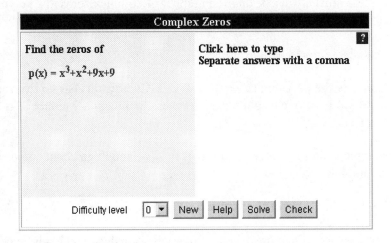

IMPORTANT: The theorem above only applies to polynomials with real number coefficients. Polynomials with nonreal coefficients are more complicated, since zeros need not appear in conjugate pairs. The examples that follow illustrate several possibilities.

$f(x) = x^2 + 1$	Has no real zeros, but does have the complex zeros i and $-i$. So here we have real coefficients and complex, nonreal zeros.
$f(x) = x^2 - 1$	Has only real zeros, 1 and -1. So here we have real coefficients and real zeros.
$f(x) = ix^2 + 1$	Has two complex zeros: $\pm \dfrac{1+i}{\sqrt{2}}$.
$f(x) = x^2 + (-1+i)x - i$	Has complex coefficients and both a real zero, 1, and a complex, nonreal zero, $-i$.

4.11 Approximating Zeros

Algebraic methods are not always able to yield zeros of a polynomial. In this case we might be satisfied with a numerical approximation to the zero. There are several methods for obtaining such approximations, but we do not discuss them here. Instead, in the following exercise set, you will see the graph of a polynomial and at least one x-intercept.

In this exercise you can see how it is possible to quickly approximate an unknown zero of a polynomial with the aid of graphs to guide the process. Try this on your calculator if it has the capabilities to graph and zoom. Some software packages have built-in routines that approximate zeros but don't normally allow you to visually observe how the algorithms are producing better and better approximations.

Hold down the mouse button near an intercept and drag the mouse from left to right. Then click on Zoom. The graph will be drawn over a smaller interval of the x-axis. When you think you have an accurate value of the zero that satisfies the requested conditions, enter it in the box provided and then click on Check. The Zoom out button will return to the previous scale. If there are any other zeros shown in the original graph, try to find them as well.

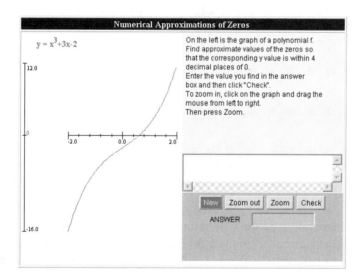

4.12 Factoring and Graphing

It's now time to put what you have learned about polynomials to use. First, practice your factoring skills using bounds on zeros, Descartes' Rule of Signs, the Rational Zeros Theorem, and synthetic division. Remember that once you've found one zero for the polynomial, you should divide the polynomial by the appropriate factor in order to continue factoring.

This exercise requires a complete factorization of the given polynomial using all your knowledge of finding zeros and factors. The factored version should be easy to graph and should be the goal of the final exercise.

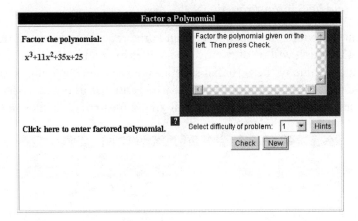

The next exercise is the ultimate goal of this entire module, where you will start with an unfactored polynomial and end up with a good graph based on its factors and zeros. Answer the questions step by step and sketch the graph on paper before displaying the graph on screen.

Now put it all together and graph the polynomials using your knowledge of factoring and graphing. In the next exercise, the first polynomials you work on will be in factored form, and then you will go on to polynomials that you will also have to factor.

In each case, once you have completed entering all the information correctly, sketch the graph of the function on paper. Then click on $\boxed{\text{Next}}$ to see the graph on the computer and check your sketch.

4.12. FACTORING AND GRAPHING

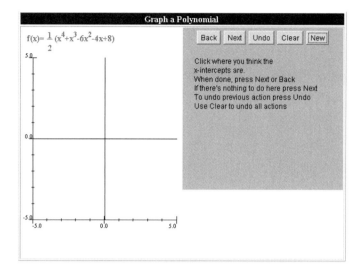

Additional Exercises

1. Factor each polynomial completely and sketch its graph, locating all intercepts.

    (a) $p(x) = \frac{1}{24}(x^3 + x^2 - 44x + 96)$

    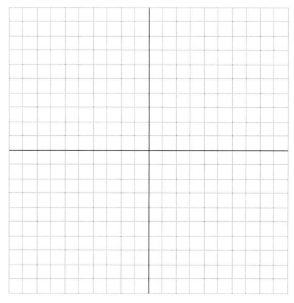

    (b) $p(x) = \frac{1}{18}(5x^3 - 26x^2 - 81x + 126)$

    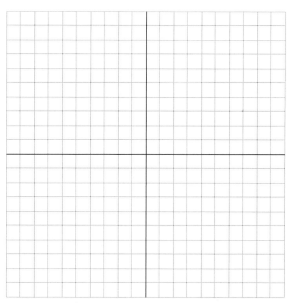

(c) $p(x) = \frac{1}{80}(6x^4 + 41x^3 - 44x^2 - 644x - 880)$

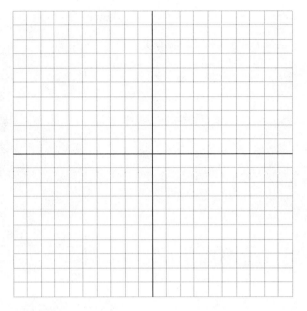

(d) $p(x) = \frac{1}{3}(x^4 - 3x^3 - 15x^2 + 19x + 30)$

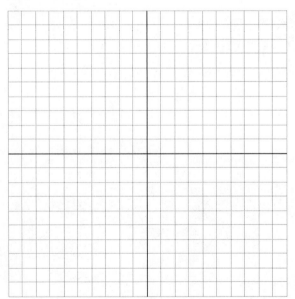

Module 5

Rational Functions

5.1 Rational Functions

So far we have discussed polynomial functions of various sorts: constant functions (e.g., $y = 2$), linear functions (e.g., $y = -3x + 4$), and higher-degree polynomial functions (e.g., $y = 2x^4 + 2x^2 + 2$ or $y = (x+1)(2x-3)$). In this module, we will study functions that are formed by taking the quotient of two polynomials. (See Section 3.1, Operations on Functions.)

> A **rational function** is one of the form $f(x) = \dfrac{g(x)}{h(x)}$, where $g(x)$ and $h(x)$ are polynomials.

A basic example of a rational function is $f(x) = 1/x$. In fact, whenever $g(x)$ and $h(x)$ are linear functions, the graph of f is a translation and/or a stretch or shrink of the graph of $y = 1/x$.

This overview of rational functions will display some of the possible behaviors we will investigate. Look in particular for the different types of rational functions and for the effects of asymptotes on the graphs.

Click below for an overview of the graphs of some rational functions.

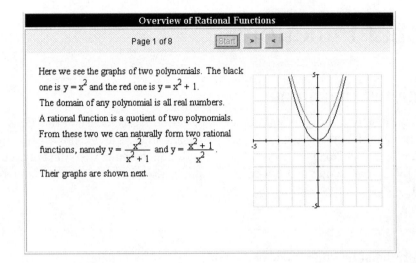

Our objective is to learn to make a rough sketch of the graph of a rational function without actually plotting a large number of points. Drawing a rational function can be complicated, and we will develop a systematic procedure for obtaining enough information to sketch the graph. We will use the four functions shown in the following table to illustrate various features found in the graphs of rational functions.

5.1. RATIONAL FUNCTIONS

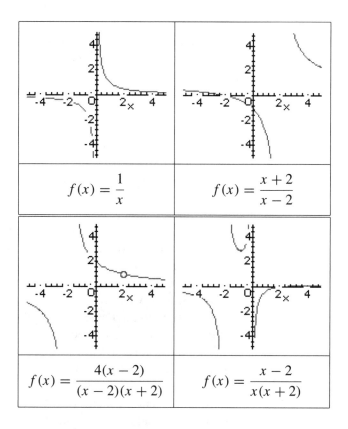

The types of information we will gather to sketch the graph of a rational function are briefly described in the following table and discussed in more detail in the following pages.

Domain and x-intercepts (Section 5.2)	Unlike a polynomial function, the domain of a rational function may not include all the real numbers. The zeros of the numerator and denominator provide essential information about the graph, so factoring will again be important.
Sign Determination (Section 5.3)	Describing where the graph lies above or below the x-axis is a great help in sketching any graph. In the discussion of polynomial functions, we found the zeros of the polynomial, used these values to partition the number line, and found the sign of the function in each of these regions. The Intermediate Value Theorem ensures that the values of $f(x)$ are either always positive or always negative in each of these regions. A rational function is not defined (and so is discontinuous) at zeros of the denominator, so we need to partition the number line using zeros of both the numerator and the denominator. Otherwise, the procedure for determining the sign of the function is the same as that discussed for polynomial functions.
Vertical Asymptotes (Section 5.4)	A rational function is not defined when the denominator is zero. If the factor that produced that zero can be canceled by an identical factor in the numerator, the graph will have a hole at that point. More commonly, if the factor cannot be removed by cancellation, the function values become huge in magnitude for values of x close to that zero. The graph becomes almost vertical, and the zero gives rise to a **vertical asymptote**.
Horizontal Asymptotes (Section 5.5)	Another important difference between graphs of rational functions and those of polynomials is the possible appearance of a horizontal asymptote. This happens in the following case: Let the values of x grow in magnitude either in the positive or in the negative direction. If the values of $f(x)$ approach a finite number c, then the graph approaches the horizontal line $y = c$. This line is called the **horizontal asymptote**.

5.1. RATIONAL FUNCTIONS

Additional Exercises

1. Is $f(x) = 0$ a rational function? Explain.

2. Is $f(x) = 2$ a rational function? Explain.

3. Is $f(x) = x^2$ a rational function? Explain.

4. Is $f(x) = \dfrac{x^2}{2^x}$ a rational function? Explain.

5. Is $f(x) = \dfrac{x^2}{x-3}$ a rational function? Explain.

6. Are the following statements true or false? If false, give a counterexample.

    (a) Every rational function has an x-intercept.

    (b) Every rational function has a y-intercept.

    (c) Every rational function has a horizontal asymptote.

    (d) Every rational function has at least one vertical asymptote.

5.2 Domain and x-intercepts

> One important difference between a rational function $f(x) = g(x)/h(x)$ and a polynomial is that f is not defined when the denominator $h(x)$ is zero. In mathematical language, we say that the domain of a rational function does not contain the zero(s) of the denominator.

This part of the investigation is all based on factoring as described in the text below.

> Factor the numerator and denominator. If a linear factor in the denominator can be canceled by an identical factor in the numerator, do so, but record the zero of this factor because it is not in the domain. This zero can produce a **hole** in the graph, as is the case for the following equation at $x = 2$:
> $$f(x) = \frac{4(x-2)}{(x-2)(x+2)}$$

> After simplifying as much as possible by cancellation, record any zeros of the numerator as **x-intercepts**. For example, look at the graph of $f(x) = (x+2)/(x-2)$ and notice the value of the x-intercept.

> Record any zeros of the denominator as the locations of **vertical asymptotes**. For example, look at the graph of $f(x) = (x+2)/(x(x-2))$ and notice the asymptotes.

Additional Exercises

1. Find the domain and x-intercept(s) of the following equation.
$$f(x) = \frac{x}{x}$$
Domain _____ x-intercept(s) _____

2. Find the domain and x-intercept(s) of the following equation.
$$f(x) = \frac{x^2 + 5x - 6}{x^2 + 1}$$
Domain _____ x-intercept(s) _____

3. Find the domain and x-intercept(s) of the following equation.
$$f(x) = \frac{x^2 + 1}{x^2 - 1}$$
Domain _____ x-intercept(s) _____

4. Find the domain and x-intercept(s) of the following equation.
$$f(x) = \frac{2x + 5}{x^2 + 2x - 8}$$
Domain _____ x-intercept(s) _____

5. Does the following equation have a removable discontinuity and, if so, at what value of x?
$$f(x) = \frac{x^2 + 1}{x - 1}$$

6. Does the following equation have a hole and, if so, at what value of x?
$$f(x) = \frac{x^2 - 2x - 3}{x^2 + 3x + 2}$$

7. Does the following equation have a hole and, if so, at what value of x?
$$f(x) = \frac{x - 1}{x^2 - 2x + 1}$$

5.3 Sign Determination

In order to sketch the graph of any function, we need to know where the function is positive and where it is negative. First, note that while rational functions are constructed from polynomials, rational functions may change sign in very different ways from the way polynomials do. Polynomial functions can change signs only as the graph crosses the x-axis. For rational functions, we may also have different signs on either side of the vertical asymptotes (see Section 5.4). That is, we must use not only the zeros of the numerator but also the zeros of the denominator to determine the possible changes of sign of a rational function.

An important procedure to help determine the sign of a rational function is the following: Find all the zeros of both the numerator and the denominator, and write them in a list in increasing order. Suppose that u and v are two adjacent values in this list with $u < v$. Then the values of the rational function are either all positive or all negative on the interval (u,v). To determine which, select any **test point** t with $u < t < v$ and evaluate $f(t)$. If $f(t) > 0$, then $f(x) > 0$ for all values of x on the interval (u,v). If $f(t) < 0$, then $f(x) < 0$ for all values of x on the interval (u,v).

Notes:

- This procedure works for all rational functions except the **zero function** defined by $Z(x) = 0$ for all x.

- If v is the first number in the list, we take u to be negative infinity. Our test number t can then be any number less than v.

- Similarly, if u is the last number in the list, we take v to be infinity. Our test number t can then be any number greater than u.

Here are illustrations of the process using two of our example functions.

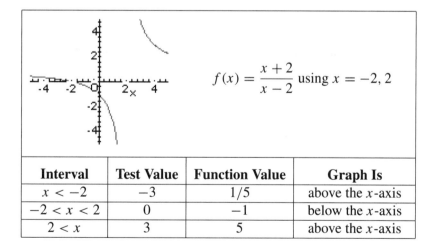

$f(x) = \dfrac{x + 2}{x - 2}$ using $x = -2, 2$

Interval	Test Value	Function Value	Graph Is
$x < -2$	-3	$1/5$	above the x-axis
$-2 < x < 2$	0	-1	below the x-axis
$2 < x$	3	5	above the x-axis

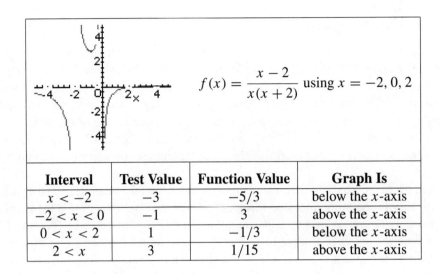

Interval	Test Value	Function Value	Graph Is
$x < -2$	-3	$-5/3$	below the x-axis
$-2 < x < 0$	-1	3	above the x-axis
$0 < x < 2$	1	$-1/3$	below the x-axis
$2 < x$	3	$1/15$	above the x-axis

Try to evaluate the function at each test value to determine the sign of the function value.

In the next set of exercises, you are given a rational function and asked to determine where the function is positive and where it is negative. Your first task is to plot on the x-axis all zeros of the numerator and denominator (if there is a common zero, plot it only once). Once you have done that, press the Next button. Then you will be asked to click in the squares where the function is positive or negative. You can then check your work by pressing the Next button once more. You will be given a chance to go back and correct any mistakes. Press the New button to get another problem.

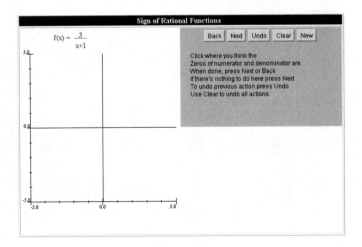

5.3. SIGN DETERMINATION

Additional Exercises

1. Find the real zeros of the numerator and denominator of the following function.
$$f(x) = \frac{x^2 + 5x - 6}{x^2 + 1}$$

   What are the intervals on which f is positive and negative?

2. Find the real zeros of the numerator and denominator of the following function.
$$f(x) = \frac{x^2 + 1}{x^2 - 1}$$

   What are the intervals on which f is positive and negative?

3. Find the real zeros of the numerator and denominator of the following function.
$$f(x) = \frac{2x + 5}{x^2 + 2x - 8}$$

   What are the intervals on which f is positive and negative?

4. Find the real zeros of the numerator and denominator of the following function.
$$f(x) = \frac{x^2 - 2x - 3}{x^2 + 3x + 2}$$

What are the intervals on which f is positive and negative?

5. Find the reals zeros of the numerator and denominator of the following function.
$$f(x) = \frac{x - 1}{x^2 - 2x + 1}$$

What are the intervals on which f is positive and negative?

6. Give an example of a rational function that is positive for all $x < 1$, negative for $1 < x < 4$, and positive for all $x > 4$.

5.4 Vertical Asymptotes

> If the numerator and denominator of a rational function f have no common factors, then we have a **vertical asymptote** at each real zero of the denominator. If they do have common factors, then one of two things can occur: If the common factor in the numerator has an exponent larger than or equal to the exponent of the common factor in the denominator, then the function has a **hole**. If the common factor in the denominator has the larger exponent, then the function has a vertical asymptote.
>
> For example, the function $f(x) = x^2/x$ has a hole at 0. Here the common factor is x, but the exponent of the common factor is larger in the numerator. The function $f(x) = x/x^2$ has a vertical asymptote at 0, since the common factor x has the larger exponent in the denominator.

Behavior about a vertical asymptote is well illustrated by the example $f(x) = 1/x$. Note that $f(x)$ is not defined at $x = 0$ but is defined for values of x as close as we want to zero. When x is very close to zero and positive, say $x = 0.0001$, then $f(x)$ is large and positive ($f(0.0001) = 10,000$). In fact, $f(x)$ can be made as large as we wish in the positive direction by making x sufficiently close to zero but still positive.

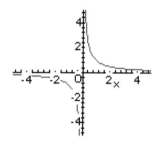

This means that when we draw the graph of $f(x) = 1/x$, the graph rises and becomes closer and closer to the vertical line $x = 0$ as x approaches zero through positive values. Symbolically we write $f(x) \longrightarrow \infty$ as $x \longrightarrow 0^+$. When x is very close to zero and negative, say $x = -0.0001$, then $f(x)$ is large in magnitude and negative ($f(-0.0001) = -10,000$). As the values of x approach zero from the left, the graph drops, becoming closer to the vertical line $x = 0$. In this case, we write $f(x) \longrightarrow -\infty$ as $x \longrightarrow 0^-$. This situation tells us that the line $x = 0$ is a **vertical asymptote** for the function $f(x) = 1/x$.

> IMPORTANT: A vertical asymptote is like an impenetrable wall—the graph of a rational function *never* crosses a vertical asymptote!

Consider again the graphs of our other three examples and how their vertical asymptotes arise.

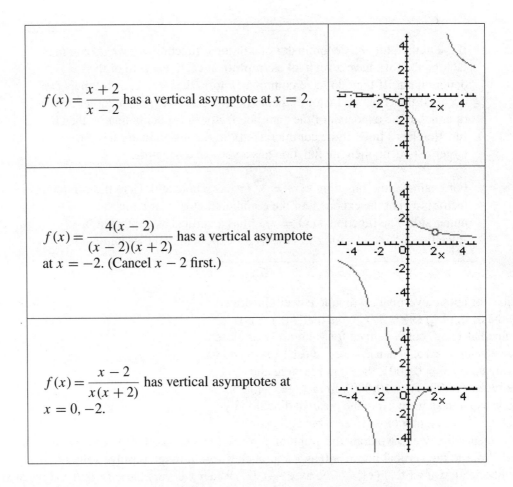

$f(x) = \dfrac{x+2}{x-2}$ has a vertical asymptote at $x = 2$.	
$f(x) = \dfrac{4(x-2)}{(x-2)(x+2)}$ has a vertical asymptote at $x = -2$. (Cancel $x - 2$ first.)	
$f(x) = \dfrac{x-2}{x(x+2)}$ has vertical asymptotes at $x = 0, -2$.	

Following is a summary of the method used to locate and indicate vertical asymptotes graphically.

> After cancelling any common factors,
>
> - Find the real zeros of the denominator. These will locate the vertical asymptotes.
>
> - Determine whether the function is positive or negative to the left and right of each asymptote. This tells whether the graph goes up or down as the graph approaches the asymptote in each direction. Note that it might go up from one side and down from the other, or both sides might be the same.

5.4. VERTICAL ASYMPTOTES

In this exercise, locate vertical asymptotes and determine the behavior of the graph on either side of the asymptote. Understanding this part of rational functions will be important later when graphing.

In this exercise, you must determine the vertical asymptote(s) of the given function. If you think that the function has a vertical asymptote at $x = 2$, for example, enter "x=2" in the box. Determine whether $f(x) \longrightarrow \infty$ or $f(x) \longrightarrow -\infty$ on the left of the asymptote, and choose the appropriate answer. Then do the same on the right side of the asymptote.

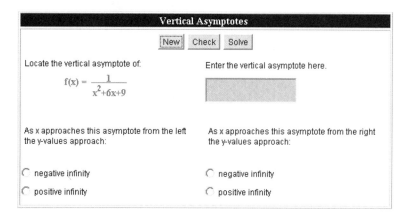

5.4. VERTICAL ASYMPTOTES

Additional Exercises

1. The function $f(x) = \dfrac{-3}{x-3}$ has a vertical asymptote at $x = 3$.

    (a) For $x < 3$, do the function values approach negative or positive infinity as x approaches 3?
    Circle your answer: ∞ $-\infty$

    (b) For $x > 3$, do the function values approach negative or positive infinity as x approaches 3?
    Circle your answer: ∞ $-\infty$

2. The function $f(x) = \dfrac{2}{x+1}$ has a vertical asymptote at $x = -1$.

    (a) For $x < -1$, do the function values approach negative or positive infinity as x approaches -1?
    Circle your answer: ∞ $-\infty$

    (b) For $x > -1$, do the function values approach negative or positive infinity as x approaches -1?
    Circle your answer: ∞ $-\infty$

3. Find the vertical asymptotes of the following function, and determine the function behavior on each side of the asymptote.

$$f(x) = \frac{x^2 + 5x - 6}{x^2 + 1}$$

4. Find the vertical asymptotes of the following function, and determine the function behavior on each side of the asymptote.

$$f(x) = \frac{x^2 + 1}{x^2 - 1}$$

5. Find the vertical asymptotes of the following function, and determine the function behavior on each side of the asymptote.

$$f(x) = \frac{2x+5}{x^2+2x-8}$$

6. Find the vertical asymptotes of the following function, and determine the function behavior on each side of the asymptote.

$$f(x) = \frac{x^2-2x-3}{x^2+3x+2}$$

7. Find the vertical asymptotes of the following function, and determine the function behavior on each side of the asymptote.

$$f(x) = \frac{x-1}{x^2-2x+1}$$

5.5 Horizontal Asymptotes

If the graph of a rational function approaches a horizontal line as the values of x assume increasingly large magnitudes, the graph is said to have a **horizontal asymptote**. This means that for very large values of x, $f(x) \approx L$. Similarly, for values of x large in magnitude but negative in sign, $f(x) \approx L$. The determination of a horizontal asymptote is fairly easy, since every rational function falls into one of three categories, as described next:

> If the degree of the numerator is less than the degree of the denominator, then the horizontal asymptote is always the x-axis; i.e., the line $y = 0$.

> If the numerator and denominator are of equal degree, the horizontal asymptote is always the ratio of the leading coefficients. So if the leading coefficient of the numerator is a and the leading coefficient of the denominator is b, then the asymptote is the line $y = \dfrac{a}{b}$.

> If the degree of the numerator is greater than the degree of the denominator, then we have no horizontal asymptote, and the y-values increase without bound as x increases in magnitude. Investigation of such graphs is outside the scope of this course. In such cases, you will get either an oblique asymptote or the graph of a higher-degree polynomial as an asymptote.

In the following table, we give four examples of rational functions. The graph of each function may be viewed by clicking on the Graph *button. The graph will then be displayed in a new window.*

Here are some examples. Click the buttons on the web site to see the graphs.

Function	Degree of Numerator	Degree of Denominator	Horizontal Asymptote	Graph
$\dfrac{x-2}{x(x+2)}$	1	2	$y = 0$	Graph
$\dfrac{x+2}{x-2}$	1	1	$y = \dfrac{1}{1} = 1$	Graph
$\dfrac{5x+3}{2x-1}$	1	1	$y = \dfrac{5}{2}$	Graph
$\dfrac{x^2+2}{x+1}$	2	1	None	Graph

An interesting case is when we have equal degrees in the numerator and the denominator, as is the case for $f(x) = \frac{5x+3}{2x-1}$. If we divide the numerator and denominator by the highest power of x (here that's the first power), we get

$$f(x) = \frac{5 + \frac{3}{x}}{2 - \frac{1}{x}}$$

As $x \longrightarrow \infty$ or $x \longrightarrow -\infty$, the values of $f(x)$ approach the number $\frac{5}{2}$, since $\frac{3}{x}$ and $\frac{1}{x}$ tend to zero when x tends to infinity.

> Suppose f is a rational function with horizontal asymptote $y = L$.
> We know that for large values of x, the value of $f(x)$ is approximately L.
> We say that f approaches the asymptote **from above** if $f(x) > L$ for all sufficiently large x. Similarly, we say f approaches the asymptote **from below** if $f(x) < L$ for all sufficiently large x.
>
> We use the same terminology for x large in magnitude but negative in sign.

In the following table we illustrate the concept of approaching the asymptote from above or below numerically. For each function in the table, we calculated the value of the function at several values in the domain. We used this to decide whether the function approached the asymptote from above or below. For simplicity, we chose functions for which the asymptote is the line $y = 0$, but this technique could be used for any horizontal asymptote.

5.5. HORIZONTAL ASYMPTOTES

Positive x

	$x = 10$	$x = 100$	$x = 1000$	Approaches asymptote
$\dfrac{1}{x^2}$	0.01	0.0001	0.000001	from above
$\dfrac{x-2}{x(x+2)}$	0.0667	0.0096	0.00099	from above

Negative x

	$x = -10$	$x = -100$	$x = -1000$	Approaches asymptote
$\dfrac{1}{x^2}$	0.01	0.0001	0.000001	from above
$\dfrac{x-2}{x(x+2)}$	−0.15	−0.0104	−0.00010	from below

In this exercise, locate the horizontal asymptotes and determine the behavior of the graph as it approaches the asymptote. Understanding this part of rational functions will be important later when graphing.

In the next exercise, find the horizontal asymptote of the given function. For example, if you think the function has the horizontal asymptote $y = 2$, enter "y=2" in the box. (Note that you are entering your answer in a text box; if your answer involves a fraction, you will need to enter the fraction in the form "a/b".) Once you have found the horizontal asymptote, determine whether $f(x)$ approaches the asymptote from above or below as $x \longrightarrow \infty$. Repeat for $x \longrightarrow -\infty$.

Horizontal Asymptotes

[New] [Check] [Solve]

Locate the horizontal asymptote of:

$$f(x) = \frac{x^2+2}{2x^2+14x+24}$$

Enter the asymptote here.

How does y approach the asymptote as x approaches negative infinity?

How does y approach the asymptote as x approaches positive infinity?

○ from above
○ from below

○ from above
○ from below

5.5. HORIZONTAL ASYMPTOTES

Additional Exercises

1. The function $f(x) = \dfrac{-3x}{x-3}$ has horizontal asymptote $y = -3$.

    (a) Does the graph of f approach the horizontal asymptote from above or below as x tends to infinity?

    (b) Does the graph of f approach the horizontal asymptote from above or below as x tends to negative infinity?

2. The function $f(x) = \dfrac{2}{x+1}$ has horizontal asymptote $y = 0$.

    (a) Does the graph of f approach the horizontal asymptote from above or below as x tends to infinity?

    (b) Does the graph of f approach the horizontal asymptote from above or below as x tends to negative infinity?

3. Find the horizontal asymptote of the following function and determine how the function approaches the horizontal asymptote as x tends to positive and negative infinity.
$$f(x) = \frac{x^2 + 5x - 6}{x^2 + 1}$$

4. Find the horizontal asymptote of the following function and determine how the function approaches the horizontal asymptote as x tends to positive and negative infinity.
$$f(x) = \frac{3x^2 + 1}{4x^2 - 1}$$

5.6 Graphing Rational Functions

We are now ready to use the information and techniques we have learned in the preceding sections to sketch the graph of a rational function. Here we've summarized the steps used in graphing a rational function based on all the information gathered. As usual, we include locating the y-intercept, if it exists.

1. Factor the numerator and denominator (into linear factors, if possible). Simplify.

2. Use the factored form of the function (including any factors you have canceled from the denominator) to determine the

   - x-intercepts (i.e., the zeros of the numerator that are not zeros of the denominator)
   - The vertical asymptotes (i.e., the zeros of the denominator that are not zeros of the numerator or zeros of the denominator that have larger multiplicity than the zeros of in the numerator)
   - The "holes" (the zeros of the numerator that have the same or larger multiplicity than the zeros of the denominator)

3. Determine where the function is positive and where it is negative.

4. Determine the behavior of the function near the vertical asymptotes.

5. Find the horizontal asymptote, if there is one, and determine the behavior of the function as $x \longrightarrow \infty$ and as $x \longrightarrow -\infty$.

6. If f is defined at 0, find the y-intercept, $f(0)$.

Now put all the techniques of this module together to graph rational functions. When stepping through the exercise, duplicate the results on paper and produce the graph on paper before revealing the graph on screen.

With all the information you have accumulated about your function through these steps, you should now be ready to sketch the graph. The next exercise will lead you through the preceding steps.

At each stage, answer the questions in the box, then press [Next]. If you are asked for an intercept, asymptote, or hole when there is none, just press [Next]. When you have completed all the questions, the software will tell you which you answered correctly and which answers need to be fixed.

If an answer needs to be corrected, return to the appropriate page (by clicking the appropriate button), press [Undo] to cancel the entries on that screen, and reenter the information.

Once all your information is correct, sketch the graph on paper. Then click on [Next] to see the graph on the computer and compare it with your sketch.

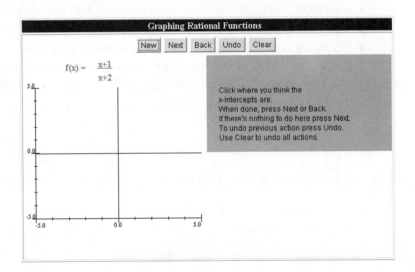

Note: The rules we have described here are helpful for all rational functions for which the degree of the denominator is not smaller than the degree of the numerator. However, these rules are not completely sufficient for functions such as the following:

$$f(x) = \frac{x^2 - 1}{x + 2} \quad \text{or} \quad f(x) = \frac{x^3 + x + 2}{(x - 1)(2x + 3)}$$

In cases like these, there may be other kinds of asymptotes: oblique straight lines, or even parabolas or higher-degree polynomials! These are beyond the scope of this course. Therefore, we have not gone into detail regarding such graphs.

5.6. GRAPHING RATIONAL FUNCTIONS

Additional Exercises

Sketch the graph of each of the six functions listed below. Use the graph paper provided. Note that you will have to choose appropriate scales for both axes. Clearly illustrate and label intercepts, holes, and asymptotes.

1. $f(x) = \dfrac{-3}{x-3}$

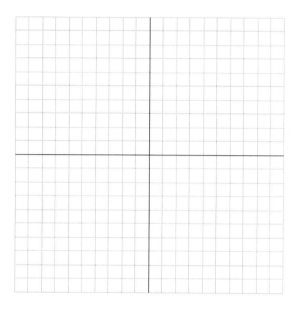

2. $f(x) = \dfrac{x^2 + 5x - 6}{x^2 + 1}$

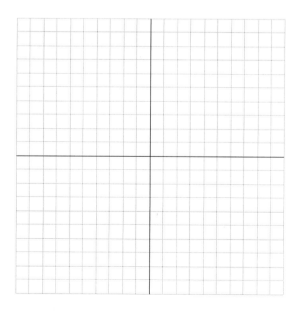

3. $f(x) = \dfrac{x^2+1}{x^2-1}$

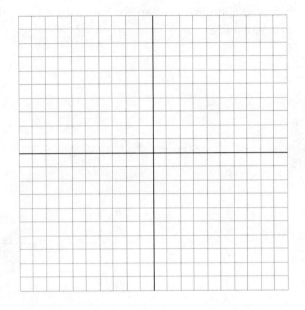

4. $f(x) = \dfrac{2x+5}{x^2+2x-8}$

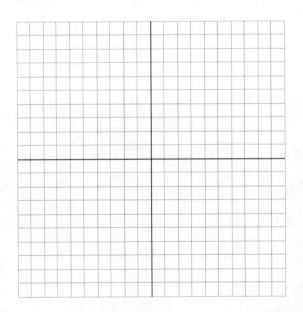

5.6. GRAPHING RATIONAL FUNCTIONS

5. $f(x) = \dfrac{x^2 - 2x - 3}{x^2 + 3x + 2}$

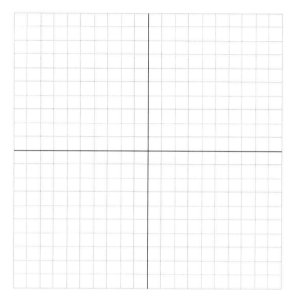

6. $f(x) = \dfrac{x - 1}{x^2 - 2x + 1}$

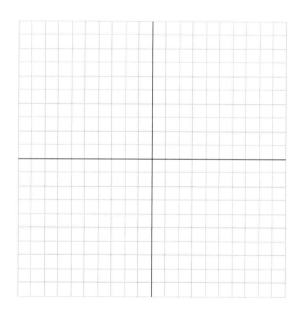

Module 6

Exponential Functions

6.1 Exponential Functions

> An **exponential function with base** a is a function of the form
> $$f(x) = a^x$$
> (a is any positive real number different from 1.)

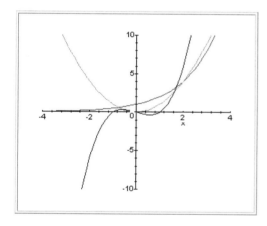

The function $f(x) = 2^x$ (its graph is in red on the web) is an exponential function. Functions of this sort are very different from polynomial functions, even though exponents appear in both types. The function $f(x) = 2^x$ is an exponential function, whereas $g(x) = x^2$ and $h(x) = x^3 - x$ are polynomial functions; see how different the graphs are from each other. In particular, note that the exponential function has a horizontal asymptote.

View this demonstration to get a comparison of the growth rates of exponential and polynomial functions. Notice that for even relatively small values of n, 2^n is much larger than n^2.

Since the preceding graphs show only a small portion of the functions, the relative sizes of $f(x) = 2^x$ and the polynomials are not evident. In fact, an exponential with base greater than 1 grows much faster than any polynomial for large values of x. To get an idea of relative sizes, this example illustrates the values of $f(n) = 2^n$ and $g(n) = n^2$ for various values of n. As you click the Next button, you will see two growing rectangles. The red rectangle is of size n^2 pixels, whereas the green rectangle is of size 2^n pixels. The value of n at each step is given at the bottom.

This is a good time to brush up on the properties of exponents. To see these, together with some examples, click on the link properties of exponents on the web.

If you are not familiar with properties of exponents, this is the time to practice them!

Exponential functions arise in a wide variety of areas in "real life," including finance, biology, physics, and many others. The base that we use often depends on the application. One in particular is the irrational number e, whose decimal value is approximately 2.718. A brief description of one way in which the number e arises follows the examples below.

6.1. EXPONENTIAL FUNCTIONS

Unrestricted population growth: A population of size $P(t)$ doubles every d (for doubling period) time units. $P_0 = P(0)$ is the initial size at time $t = 0$.	**Example:** A population starts with 400 individuals and doubles every 5 years.
$P(t) = P_0(2)^{t/d}$	$P(t) = 400(2)^{t/5}$

Radioactive decay: An amount $A(t)$ of a radioactive material decays by losing half its mass every h (for half-life) time units. $A_0 = A(0)$ is the initial mass at time $t = 0$.	**Example:** An initial amount of 750 g of a material with a half-life of 12 days.
$A(t) = A_0\left(\frac{1}{2}\right)^{t/h}$ $= A_0(2)^{-t/h}$	$A(t) = 750\left(\frac{1}{2}\right)^{t/12}$ $= 750(2)^{-t/12}$

Compound interest: The amount of money $A(t)$ in a compound interest account from principal P at interest rate r (expressed as a decimal) compounded n times per year invested for t years.	**Example:** Invest $600 at 6% compounded monthly.
$A(t) = P\left(1 + \dfrac{r}{n}\right)^{nt}$	$A(t) = 600\left(1 + \dfrac{0.06}{12}\right)^{12t}$

INTERACTIVE COLLEGE ALGEBRA: A WEB-BASED COURSE © 2005 Key College Publishing

Continuously compounded interest: The amount of money $A(t)$ in a continuously compounded interest account from principal P at interest rate r (expressed as a decimal) invested for t years.	**Example:** Invest \$320 at 4.25% compounded continuously.
$A(t) = Pe^{rt}$	$A(t) = 320e^{0.0425t}$

Sometimes we manipulate these formulas to fit a special need. For example, if we consider a credit card bill as the credit company investing in us instead of us investing in the bank, then we must pay the interest out to them. One way to do this is to use daily compounding on the average daily balance. We can get the following formula for a monthly bill by noting that a month with d days contains $t = d/365$ of a year, so that $nt = d$.

Credit card interest: The balance due after a month containing d days with average daily balance B_0 at interest rate r (expressed as a decimal).	**Example:** Balance due after 31 days with an average daily balance of \$870 at 19%.
$B = B_0\left(1 + \dfrac{r}{365}\right)^d$	$B = 870\left(1 + \dfrac{0.19}{365}\right)^{31}$

Investigating the ways in which money increases via compound interest shows how the interesting and important (irrational) number e arises. Suppose you invest \$1 at a compound interest rate of 100% compounded n times per year for one year. How much money will you have after one year? Of course the answer will depend on the value of n. The formula that tells you how much money you have is given in the following activity. See how your money grows over just one year by moving the scroll bar to the right.

6.1. EXPONENTIAL FUNCTIONS

You should now view this demonstration that approximates the number e by using the compound interest formula.

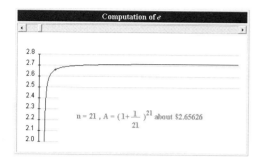

Move the indicator on the top scroll bar to increase the value of n. As n increases, the value of A gets closer to e. You never actually reach the number e this way, and the formula will always produce a number less than e. The true value of e could not be given by such a formula involving fractions and integer powers since e is an irrational number.

When we use the number e as the base for an exponential function, we call it the **natural exponential function**.

You should now try this exercise that utilizes the exponential formulas described earlier. The applications include population growth, radioactive decay, and interest problems. A calculator is required to approximate the answers.

Now is the time to try out your skills on some problems using exponential functions. Click on Radioactive, Population, or Interest to choose the type of question you want to work on.

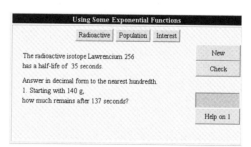

As for functions of any other sort, we need to be able to graph exponential functions, which we will learn in the next section.

INTERACTIVE COLLEGE ALGEBRA: A WEB-BASED COURSE © 2005 Key College Publishing

6.1. EXPONENTIAL FUNCTIONS

Additional Exercises

1. Find the amount of a radioactive material remaining, given its half-life h, initial amount Q_0, and elapsed time t.

    (a) $h = 21$ min, $Q_0 = 9$ g, and $t = 54$ min

    (b) $h = 3$ yr, $Q_0 = 250$ g, and $t = 9$ months

    (c) $h = 1200$ yr, $Q_0 = 2000$ g, and $t = 5000$ yr

2. If a population grows exponentially, find the size of the population, given its doubling period d, initial size P_0, and elapsed time t.

    (a) $d = 10$ yr, $P_0 = 400$ people, and $t = 86$ yr

    (b) $d = 3$ days, $P_0 = 10000$ people, and $t = 2$ weeks

    (c) $d = 50$ yr, $P_0 = 3000$ people, and $t = 65$ yr

3. Find the balance in a savings account, given the initial principal P, interest rate r and method of compounding, and elapsed time t.

    (a) $P = \$700$, $r = 6\%$ compounded quarterly, and $t = 10$ yr

    (b) $P = \$700$, $r = 6\%$ compounded continuously, and $t = 10$ yr

    (c) $P = \$300$, $r = 8\%$ compounded daily, and $t = 30$ yr

    (d) $P = \$300$, $r = 8\%$ compounded continuously, and $t = 30$ yr

6.2 Graphing

The graphs of **exponential functions** are not very different from one another; in this section we will see how to graph any particular exponential function. Our main tools for this are knowledge of the shapes of the basic exponential functions, and the graphing techniques discussed in Section 2.2.

You should go through this overview to get basic information about graphs of exponential functions. Look for information on domain, range, shapes of graphs, different bases, and comparisons to other types of graphs.

First, read the following overview of the exponential functions and their graphs.

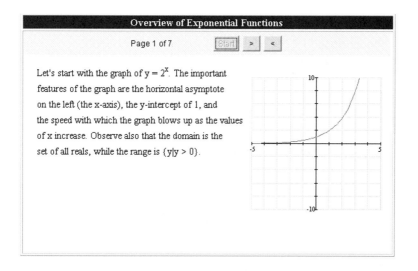

Now you can practice graphing exponential functions in two ways. The first set of exercises gives you the basic shape and asks you to use graphing techniques to adjust it to the specific function you are given.

Apply graphing techniques such as shifting and scaling to the new class of exponential functions. You will find out how a vertical shift affects the range of an exponential function. Sketch the graphs on paper, too, to practice graphing skills.

First select one of the three basic types of exponential functions. You will see the general shape of the type of exponential function you have chosen. Now, use the controls to transform this graph to the graph of the given function. When done, click on Check it , or, if you have trouble, Graph it .

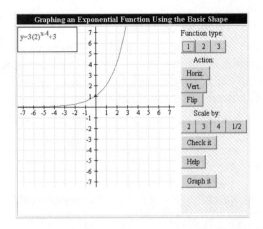

Now try this exercise in which your knowledge of the behavior of the graph is entered step by step before any graph is displayed. These, too, should be graphed on paper before reviewing on screen.

In the next set of exercises, you are asked to graph exponential functions "from scratch." Answer each of the questions on the screen and after each one, click Next . When you have completed all the questions, you will be told how well you did and you will have the opportunity to correct any mistakes. Once you have no more errors, sketch the graph in your notebook. Then click on Next to see the graph of the function and compare it with the graph you sketched.

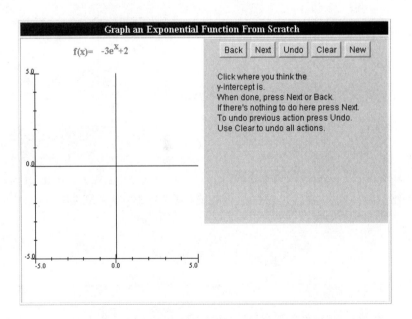

6.2. GRAPHING

Additional Exercises

Sketch the graph of each of the following exponential functions on the graph space provided. Clearly label the y-intercept and the location of the horizontal asymptote. Note that you will need to decide appropriate scales for both axes.

1. $f(x) = 2e^x - 3$

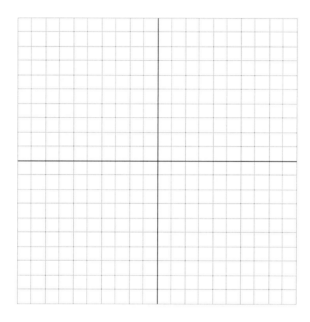

2. $f(x) = -3e^{x/2} + 1$

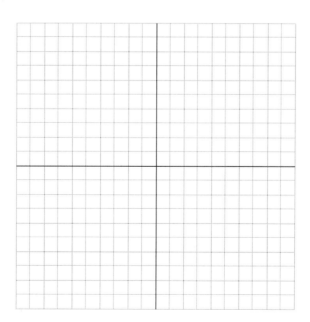

3. $f(x) = e^{-x} + 2$

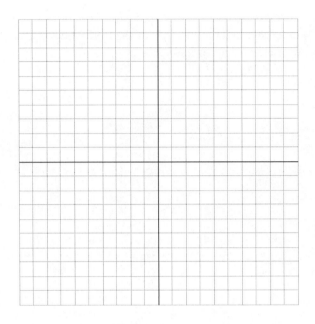

4. $f(x) = \frac{1}{2}e^{x-1} - 2$

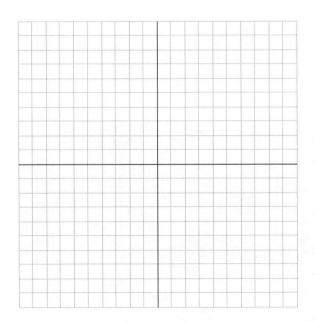

6.2. GRAPHING

5. $f(x) = 2^{-0.3x} - 1$

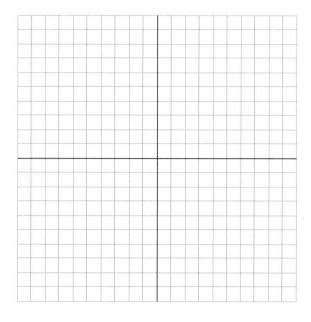

6. $f(x) = -2^{0.9x}$

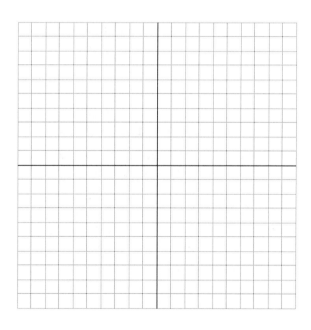

7. $f(x) = 3(2^{0.4x-2})$

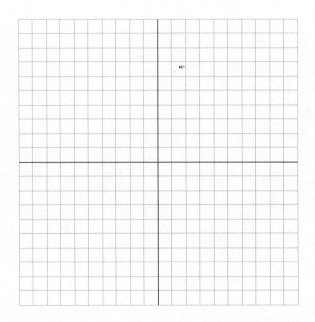

8. An exercise in Section 4.1 gave a polynomial that approximates the function $y = \sqrt{x+1}$ for values of x close to zero. We can do the same for $f(x) = e^x$. Using methods of calculus, it can be shown that the polynomial $p(x) = \frac{1}{6}x^3 + \frac{1}{2}x^2 + x + 1$ is close in value to $f(x)$ for x close to zero.

   (a) Verify that $p(0) = f(0)$. Thus, for $x = 0$ the error in approximating $f(0)$ by $p(0)$ is zero. Another way to write this is $f(0) - p(0) = 0$.

   (b) Estimate the error for $x = 0.1$ this way. Use a calculator to find both $f(0.1)$ and $p(0.1)$ accurate to seven decimal places. Then find the difference $f(0.1) - p(0.1)$.

   (c) Estimate the error for $x = 0.2$ in a similar manner. Use a calculator to find both $f(0.2)$ and $p(0.2)$ accurate to seven decimal places. Then find the difference $f(0.2) - p(0.2)$.

   (d) Since $x = 3$ isn't nearly as close to zero as 0.1 or 0.2, we might expect less accuracy in our estimation. Use a calculator to find both $f(3)$ and $p(3)$ accurate to seven decimal places. Then find the difference $f(3) - p(3)$.

   (e) Graph both $f(x)$ and $p(x)$ together on a graphing calculator or using graphing software. Choose an interval that contains the number zero. For what x-values does $p(x)$ appear to be a very good approximation to $f(x)$?

Module 7

Logarithmic Functions

7.1 Introduction to Logarithms

We have already seen and worked with functions of the form $f(x) = a^x$, the exponential function with base a, for $0 < a < 1$ or $1 < a$. For each a, the resulting function is one-to-one and therefore has an inverse. Note that just as different values of a in exponential functions give different functions, so do different values of a as the base of a logarithmic function give different log functions.

> The inverse of the exponential function $f(x) = a^x$ is called a **logarithmic function with base a** and is written $\log_a x$ or $\log_a(x)$.

Before we go into a detailed discussion of logarithmic functions, let us consider the following relationship between numbers a, b, and c:

> $c = \log_a b$ if and only if $a^c = b$.

In order to get used to this relationship, study these first examples and think about the role of a, b, and c in each statement:

$2 = \log_3 9$	because	$3^2 = 9$
$5 = \log_2 32$	because	$2^5 = 32$
$-3 = \log_{10}(0.001)$	because	$10^{-3} = 0.001$
$4 = \log_e(e^4)$	because	$e^4 = e^4$

Let's apply the technique developed in Section 3.2 to find a formula for the inverse of an exponential function. If we are given the exponential $y = f(x) = 2^x$, how do we find the inverse formula?

1. Switch x and y to get $x = 2^y$.

2. Solve for y to get $y = f^{-1}(x)$.

How do we solve $x = 2^y$ for y? That's exactly what the statement $y = \log_2 x$ means by definition. One of the important uses of logs will then be to solve exponential equations (in which the variable is located in an exponent).

It is critical to understand the conversion process between statements in exponential and logarithmic forms, because this process underlies the use and manipulation of logarithms. Practice this next exercise until the switch becomes automatic.

Now we should practice translating logs to exponents and back. Note that we are looking for a certain form of an answer. For example, if the question asks you to rewrite "$\log_2 x = 3$," your answer should be typed in as "$x=2^3$" and not the simplified answer of "$x=8$" (which would be reported as incorrect). Also, don't forget to parenthesize expressions under the log. For example, the logarithmic form of $5^x = y$ is $x = \log_5 y$, and you can enter your answer as "$\log_5 y$." However, the logarithmic form of $8^x = y - 8$ is $x = \log_8(y - 8)$, and you need to enter the answer as "$\log_8(y-8)$," not "$\log_8 y-8$."

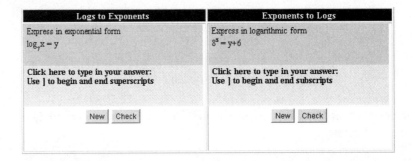

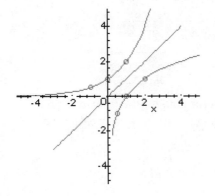

As usual, we obtain the graph of an inverse by reflecting the original graph over the line $y = x$. Here we have the graph of $y = 2^x$ (in green on the web) and its inverse $y = \log_2 x$ (in red on the web). Since the basic exponential functions have domain all real numbers and range $y > 0$, we know that the basic logarithmic functions must have domain $x > 0$ and range all reals. This is important to remember.

7.1. INTRODUCTION TO LOGARITHMS

> The domain value you "plug into" the log must be positive, but the value of the resulting log can be any real number: positive, negative, or 0.

Based just on the definition of logarithms, we have four properties to remember:

$\log_a 1 = 0$	$\log_a a = 1$	$\log_a(a^x) = x$	$a^{(\log_a x)} = x$

Take a moment to figure out why the first two hold, using only the relationship between logarithmic and exponential statements. For the last two, remember some familiar statements that hold for any pair of functions that are each the inverse of the other: $f^{-1}(f(x)) = x$ and $f(f^{-1}(x)) = x$.

Note: The base of a logarithmic function is an essential part of this function. For this reason, the next section, 7.2, Base 10 and Base e, discusses logarithm bases. At the end of that section, you will find graphing exercises for logarithmic functions with base 10 and base e.

This is a good time to go to the web and scroll through the following overview of logarithmic functions. This will give you a glimpse of some of the graphing aspects we will cover. For now, just read through each step to recall how inverse graphs are obtained by reflection, and become familiar with their basic shape.

For an overview of graphing, you can run through the following examples before proceeding to the next section.

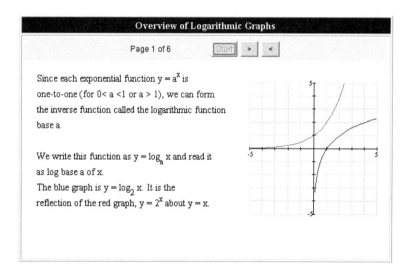

INTERACTIVE COLLEGE ALGEBRA: A WEB-BASED COURSE © 2005 Key College Publishing

Additional Exercises

1. What is the base in $3^x = y$?   Base = _____
   Rewrite the equation in logarithmic form, with the logarithm having the base you identified.

2. Rewrite $2^8 = 256$ in logarithmic form.

3. Express each of the following equations in exponential form.

   (a) $\ln 4 = z$

   (b) $\log_3 61 = w$

   (c) $\log_x 12 = 5$

   (d) $\log_4 x = -3$

   (e) $\log_3 z = -2$

4. Express each of the following equations in logarithmic form.

   (a) $2^4 = w$

   (b) $a^4 = 7$

   (c) $3^x = 11$

   (d) $2^4 = w + 1$

   (e) $3^{-x} = 11$

5. What is wrong with the following statement.
   $\log_{-3} 9 = 2$ in exponential form is written as $(-3)^2 = 9$.

7.2 Base 10 and Base e

A few values of logarithms can be found immediately just by knowing the definition of logarithm. For example, we know that $\log_2 8 = 3$ because we know that $8 = 2^3$. But suppose we start with $2^y = 73$. That translates to $y = \log_2 73$, but what is that number, and how do we approximate it? What about $y = \log_4 73$? These will be different from each other, since in the first case we have a logarithm base 2 and in the second case, a logarithm base 4. Remember: The base is an essential part of the logarithmic function!

Most calculators have two log functions built in: base 10, which is called the **common logarithm**, and base e, which is called the **natural logarithm**. (The buttons on your calculator are most likely labeled "log" and "ln" for these two functions.) Since these two are used quite extensively, each has its own special notation, as follows:

Name	Notation	Meaning
Common logarithm	$\log x$	$\log_{10} x$
Natural logarithm	$\ln x$	$\log_e x$

You should figure out now how to use these two functions on your calculator. Try it with the following values to see if you get the same results. To four decimal places, $\log(2371) = 3.3749$ and $\ln(2371) = 7.7711$.

One obvious question is, How are we going to get a value for $y = \log_2 73$ if we don't have a base 2 button on our calculator? It turns out to be fairly easy to write a log of one base in terms of logs of any other base we choose; so to use a calculator, pick base 10 or base e as your second base. Here's how it works if we want to use base 10:

If $73 = 2^y$,

then $73 = (10^{\log 2})^y = 10^{y \log 2}$.

So, $y \log 2 = \log 73$.

Finally, $y = \dfrac{\log 73}{\log 2} = 6.1898$.

The value 6.1898 is, of course, obtained from a calculator and rounded off. Do you get the same value using base e? Try it and see. That is, follow the same computation as above, but instead of log base 10, use ln, the natural log, and instead of 10 as the base in the second step, use e. You should get the same result, since after all, $\log_2$ is a function, and a function can have only one result when you plug in a given number.

Using this same technique, we can convert from any base to any other base. The general formula for this conversion is called the **change of base formula**:

$$\log_a x = \frac{\log_b x}{\log_b a}$$

Think of a as the base you started with and b as the base you want to use. (Remember that b can be any positive number that is not 1.) Thus, this formula says, for example, that

$$\log_5 3680 = \frac{\log_3 3680}{\log_3 5} = \frac{\log_7 3680}{\log_7 5} = \frac{\log 3680}{\log 5} = \frac{\ln 3680}{\ln 5} = 5.1016$$

One of the reasons base 10 is commonly used is simply that it "fits" our decimal system for writing numbers. Try these values on your calculator:

$$\log 4.32 = 0.6355$$
$$\log 43.2 = 1.6355$$
$$\log 432 = 2.6355$$
$$\log 4320 = 3.6355$$

As you can see, in the common log the digits determine the fractional part of the log, while the placement of the decimal point determines the integer part. Guess the value of log (4,320,000), and check your guess on your calculator. If you think about exponents for a moment, you can see that if $10^{0.6355} = 4.32$, then certainly $10^{3.6355} = (10^3)(10^{0.6355}) = (1000)(4.32) = 4320$.

Practice sketching the graphs of shifted and scaled log functions using the techniques learned in Module 2, Functions and Graphing. Try both base 10 and base e. As usual, try sketching the graph on paper as well as manipulating the graph on screen.

The usual graphing techniques of shifting and scaling can be applied to log functions, as this next exercise set illustrates. Your job is to obtain the graph of the displayed function by manipulating the initial graph. You may choose to graph log base 10 or log base e.

7.2. BASE 10 AND BASE e

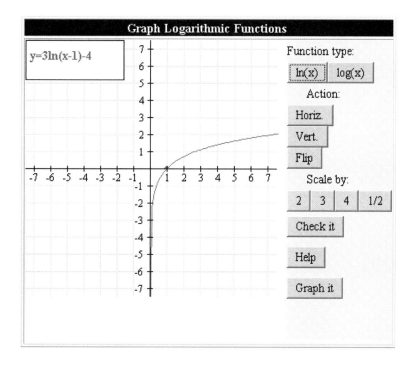

Additional Exercises

1. Write $\log_2 x$ in terms of $\ln x$.

2. Write $\log_3 x$ in terms of $\log x$.

3. Use a calculator to find

    (a) $\log 12$

    (b) $\log 120$

    (c) $\log 1200$

    (d) $\log_3 61$

    (e) $\log_2 34$

    (f) $\ln 10$

4. Sketch the graph of each of the following functions. Clearly illustrate and label intercepts and asymptotes.

    (a) $y = \log x + 2$

    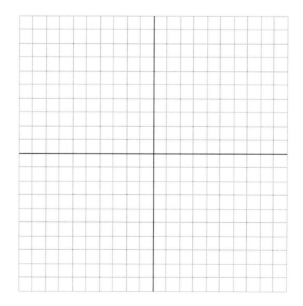

(b) $y = \log(x+1) - 10$

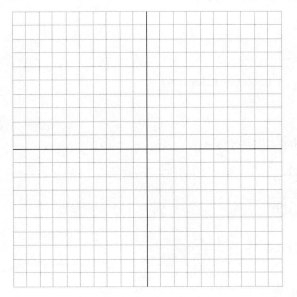

(c) $y = -\ln(x-4) + 1$

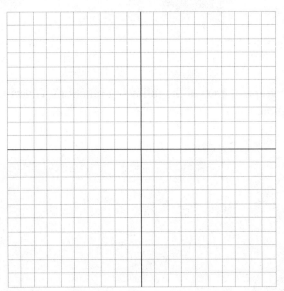

(d) $y = -3\ln x - 2$

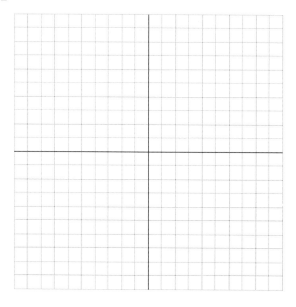

7.3 Inverses

Switching between log and exponential forms allows us to find inverses of functions that are translated and scaled versions of the basic logarithmic and exponential functions. This next problem set requires you to enter a formula for the inverse of the displayed function. Here's an example:

Start with this equation:
$$y = 3 + (5)(4^{9x})$$

Switch variables:
$$x = 3 + (5)(4^{9y})$$

Solve for y:
$$\frac{x-3}{5} = 4^{9y}$$
$$\log_4\left(\frac{x-3}{5}\right) = 9y$$
$$y = \frac{1}{9}\log_4\left(\frac{x-3}{5}\right)$$

It is important to know how to find an inverse formula for a given log or exponential function. This next exercise displays both the original and inverse graphs to emphasize the relationship between the two. Practice finding inverse formulas to become familiar with the steps and utilize the conversion step between log and exponential forms.

The following problems all use base e. The algebraic steps you use could, of course, be used on any base simply by choosing the corresponding log, as shown in the preceding example.

MODULE 7. LOGARITHMIC FUNCTIONS

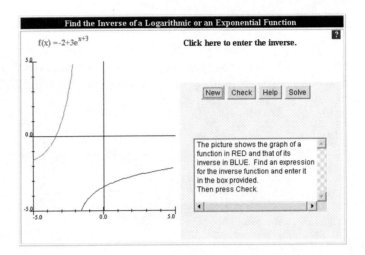

Additional Exercises

1. Find the inverse function of f given by $f(x) = e^{2x}$. Determine the domain of the inverse function. Sketch the graphs of both f and its inverse.

$f^{-1}(x) = $ _____

Domain of f^{-1}: _____

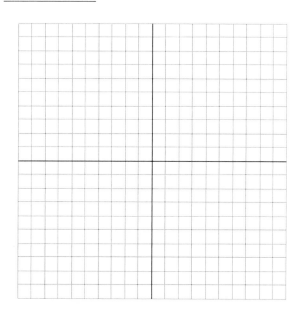

2. Find the inverse function of f given by $f(x) = 3e^{2x} + 4$. Determine the domain of the inverse function. Sketch the graphs of both f and its inverse.

$f^{-1}(x) = $ _____

Domain of f^{-1}: _____

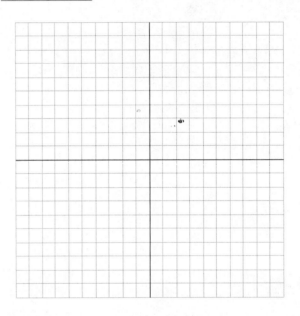

7.3. INVERSES

3. Sketch the graph of $y = e^{x^2}$. Does this function have an inverse? Explain.

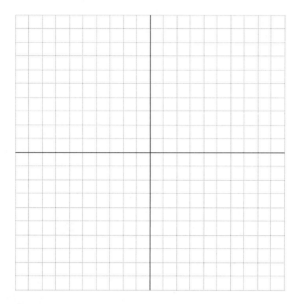

4. Sketch the graph of $y = \ln|x|$. Does this function have an inverse? Explain.

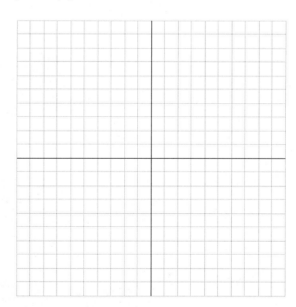

7.4 Properties of Logarithms

The real power of using logarithms lies in the properties (or rules) presented here. These are actually logarithmic analogs of well-known properties of exponents. For $x, y > 0$, and any real number r, we have the following:

Rule Number	Logarithmic Property	Exponential Analog
1	$\log_a(xy) = \log_a x + \log_a y$	$a^m a^n = a^{m+n}$
2	$\log_a(x/y) = \log_a x - \log_a y$	$a^m / a^n = a^{m-n}$
3	$\log_a(x^r) = r \log_a x$	$(a^m)^n = a^{mn}$

The rules are sometimes described by saying "logarithms turn multiplication, division, and exponentiation into addition, subtraction, and multiplication," respectively. We will see that these rules form the basis for solving exponential equations, so they should be understood and mastered.

On the web page, click on the link arithmetic with logarithms *to see an example of how logarithms were used to perform difficult hand calculations prior to the invention of calculators. The example will open up in a new window.*

Before the production of inexpensive calculators, logarithms afforded an extremely efficient way to carry out calculations with large numbers. For an example of how this was done, click on the link arithmetic with logarithms.

A good way to practice these rules is to rewrite complicated logs by expanding them in terms of simpler logs. Follow the steps in the next example to see which rules are applied.

$$\log\left(\frac{x^2 w^4}{y^3 z^7}\right) = \log(x^2 w^4) - \log(y^3 z^7)$$
$$= \log(x^2) + \log(w^4) - (\log(y^3) + \log(z^7))$$
$$= 2\log x + 4\log w - 3\log y - 7\log z$$

As you can see, there is a nice pattern here. The factors in the numerator correspond to the log terms added in, and the factors in the denominator correspond to the log terms subtracted. Notice the exponent sits out in front in each case, due to rule 3. Try some log simplifications in the next exercise set.

Practice simplifying the logarithms by expanding into sums and differences of logarithms of the individual variables. Follow the steps in the example above, and recall that $\log(1) = 0$.

$$\log\left(\frac{x^7 y^2}{z^4 w^2}\right) =$$

7.4. PROPERTIES OF LOGARITHMS

Additional Exercises

1. Express $\log 3 + \log 5$ as the logarithm of one number.

2. Express $\log 3 + \log x$ as the logarithm of one expression.

3. Express $2 \log 3 - \log x$ as the logarithm of one expression.

4. Write $\ln\left(\dfrac{2x}{3x^2}\right)$ as a sum or difference of logarithms.

5. Write $\ln\left(\dfrac{3x^5 z^3}{2y^2}\right)$ as a sum or difference of logarithms.

6. What is wrong with the following argument?

$\log(x+1) = \log x + \log 1$	(by the properties of logs)
$\log(x+1) = \log x$	(since $\log 1 = 0$)
$\log(x+1) - \log x = 0$	(subtract $\log x$ from both sides)
$\log \dfrac{x+1}{x} = 0$	(by properties of logs)
$\dfrac{x+1}{x} = 1$	(since $\log a = 0$ if and only if $a = 1$)
$x + 1 = x$	(clearing fractions)
$1 = 0$	(subtracting x from both sides)

7.5 Logarithmic and Exponential Equations

Logarithmic Equations

There are two types of very basic logarithmic equations that can be solved using only the definition of logarithms and the relationship between logarithms and exponents (see Section 7.1, Introduction to Logarithms). For these equations, we sometimes use a calculator to approximate powers.

The first type of equation that we will consider has the following form:

| $\log_a u = \log_a v$ | Two logs with the same base, set equal to each other. |

Example: $\log_6(5x + 2) = \log_6(2x + 11)$

Solution: Since log base 6 is one-to-one (remember that it has an inverse), these two are exactly equal when $5x + 2 = 2x + 11$, that is, when $x = 3$. Remember: Always try your solution in the original to make sure that the resulting values really are in the domain of the log. Here we get $(5)(3) + 2 = (2)(3) + 11 = 17 > 0$, so this is indeed a solution of the original equation.

This method of solving this type of equation is independent of the base of the logarithm. Thus, solving the equation $\ln(5x + 2) = \ln(2x + 11)$ would employ the same method, and the solution would be identical.

In the next exercise, after rewriting each equation as one log equal to another, the problem will reduce to an equation that contains no logarithms. Use the example above as a guide.

In this exercise, you are required to solve this type of logarithmic equation. You may have to use some of the properties of logarithms first to put the equation in the required form. However, all the equations here can be solved without the use of a calculator. Remember to check your solutions in the original equation. Primarily, in solving these equations, you need to recall that the domain of logarithm functions is positive real numbers.

If a problem requires more than one solution, separate your answers using a comma, colon, or similar character. A space *cannot* be used to separate multiple answers.

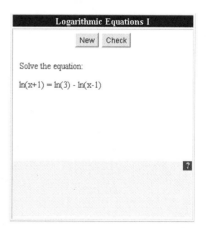

The second type of logarithmic equation we consider has the following form:

| $\log_a u = b$ | A log set equal to a fixed number. |

Example: $\log_5(4x - 2) = 3$

Solution: We know, by the definition of log base 5, that this equation is equivalent to $4x - 2 = 5^3$. So, $4x - 2 = 125$, and $x = 127/4$.

Example: $\log_2(x^2 - 1) = 5.7$

Solution: Here's where a calculator can approximate answers for us if we like. This statement is equivalent to $x^2 - 1 = 2^{5.7}$, so $x = \pm(2^{5.7} + 1)^{1/2}$. By calculator (try it!), we see that $x \approx \pm 7.279$.

Solve this type of logarithmic equation in the next exercise. After rewriting each equation in exponential form to eliminate the logarithm, solve for x and use a calculator if necessary to approximate answers. Use the example above as a guide.

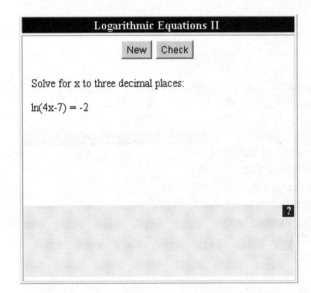

Here is an exercise on solving the second type of logarithmic equation. This time you will need to use a calculator. Remember that "log" means logarithm base 10, and "ln" means logarithm base e.

Exponential Equations

> An equation that has a variable in an exponent is called an **exponential equation**.

There are many types of exponential equations, but we shall only consider two basic types here. The first type is of the form $a^b = a^c$, and its solution relies on the fact that exponential functions are one-to-one.

Example: Solve $2^{4x-4} = 2^{x-1}$ for x.

Solution: Since the exponential function with base 2 is one-to-one, the two expressions are equal only if $4x - 4 = x - 1$, or $x = 1$.

Note, however, that we could disguise this equation in the following way: Since $2^2 = 4$, then $2^{4x-4} = 2^{2(2x-2)} = 4^{2x-2}$, and the problem could have been restated, with the same solution, as "Solve $4^{2x-2} = 2^{x-1}$ for x."

Solve this type of exponential equation in the next exercise. After rewriting each equation using the same base on both sides, if necessary, set the exponents equal to each other and solve for x. Use the example above as a guide.

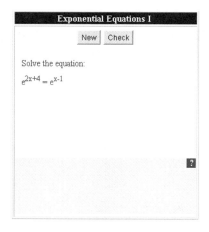

You are asked to solve this type of exponential equation in this exercise. The method of solving is as we just described—that is, use the fact that exponential functions are one-to-one.

There should be no need to use a calculator in this exercise. All answers are integers or rational numbers. If a problem requires more than one solution, separate your answers using a comma, colon, or similar character. A space *cannot* be used to separate multiple answers.

The second type of equation we consider is of the form $a^b = c$, where b will be an expression involving a variable. This type of equation is often solved, if it can be solved, as follows: Since a^b and c are equal, it follows that $\log(a^b) = \log c$. Apply rule 3 from Section 7.4 to the lefthand expression to get $b \log a = \log c$, and continue from there. The simplest type of such an equation would be something like the following:

Example: Solve $8^x = 573$ for x.

Solution: Notice that the form of this equation is an exponential set equal to a constant. That's exactly the form you want, since this way rule 3 can immediately get access to the variable.

$$8^x = 573$$
$$\log(8^x) = \log 573$$
$$x \log 8 = \log 573$$
$$x = \frac{\log 573}{\log 8} \approx 3.0541$$

As you can see, once the variable is brought down as a factor, it's a matter of simple algebraic steps to isolate the variable. Try some of these practice problems next.

Solve this type of exponential equation in the next exercise using the example above as a guide. The same three steps should work on every problem here.

Exponential Equations II

Solve for x to three decimal places:

$3^{6x} = 17625$

7.5. LOGARITHMIC AND EXPONENTIAL EQUATIONS

Of course, not every exponential equation starts out by being as simple as these. The formulas we used to describe such things as radioactive decay and compound interest are slightly more complicated and require a bit of manipulation before applying this technique. Consider a radioactive material with a half-life of 15 years. If we start with 300 g of material, the formula we use to describe the decay is

$$A = (300)(2^{-t/15})$$

If we ask when there will be 117 g of radioactive material left, that is, when A will equal 117, we are calling for a solution to this equation:

$$117 = (300)(2^{-t/15})$$

To solve this equation, isolate the exponential first by dividing throughout by 300, and then use the same steps you used above:

$$\frac{117}{300} = 2^{-t/15}$$

$$\log\left(\frac{117}{300}\right) = \log(2^{-t/15})$$

$$\log\left(\frac{117}{300}\right) = -\frac{t}{15}\log 2$$

$$t = -15\left(\frac{\log(\frac{117}{300})}{\log 2}\right) \approx 20.3768$$

Of course you can use natural log instead of common log and get the same answer. Following are some problems that use a variety of exponential formulas that you have seen. You must come up with the correct formula first. The questions are in two parts: part one involves using your calculator to evaluate your function, and part two requires that you solve an exponential equation. Give them a try.

Solve these exponential word problems using our standard formulas for exponential growth, exponential decay, and compound interest. Use the decay example above as a guide. Use a calculator to approximate answers, but do not round numbers before the final answer is calculated to avoid what could become severe round-off error. (For example, in a compound interest problem, if $1 + r/n = 1.004875664$, rounding to 1.005 before finishing the calculation would produce considerable error.)

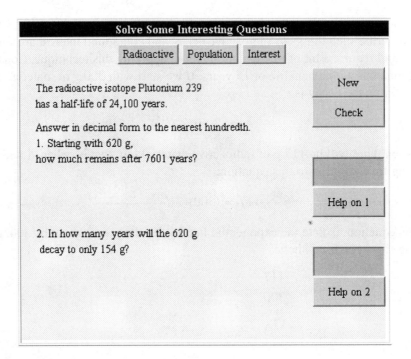

Additional Exercises

1. Solve $e^{3x} = e^{x-2}$ for x.

2. Solve $\log x + \log 3 = \log(x - 2)$.

3. Use a calculator to solve $2^x = 7$.

4. Use a calculator to solve $\ln x = 4$.

5. Solve $3^x = 15$.

6. Solve $3^{x^2} = 15$.

7. Solve $\ln x + \ln(x + 4) = \ln 9 - \ln 4$.

8. Solve $\log_3 x = 4$.

9. Solve $\log(4x + 17) = -1$.

10. Solve $e^{x-4} = \dfrac{1}{e^{x+5}}$.

11. Newton's law of cooling states that the temperature difference between an object and its environment at time t is Ae^{-kt}, where k is a constant and A is the temperature difference between the object and its environment at $t = 0$. A cup of coffee having an initial temperature of 80° Celsius, is in a room at a constant temperature of 25° Celsius. After one hour, the temperature of the coffee is 32° Celsius.

    (a) Write temperature T as a function of time t.
    (b) Use the information given to find the value of k.
    (c) How long did it take for the coffee to reach 40° Celsius?

12. The radioactive isotope Protactinium-231 (^{231}Pa) has a half-life of 32,700 years.

    (a) Starting with 670 g, how much ^{231}Pa remains after 100,000 years?

    (b) In how many years will the 670 g decay to only 200 g?

13. The radioactive isotope Curium 247 (^{247}Cm) has a half-life of 16 million years.

    (a) Starting with 45 g, how much ^{247}Cm remains after 40 million years?

    (b) In how many million years will the 45 g decay to 3 g?

14. A population of spiders doubles in size every three years.

    (a) Starting with a population of 300 spiders, find the size of the population after 11 years.

    (b) In how many years will the population reach 5000?

15. Suppose we invest $2000 in a savings account that pays 4% compounded daily.

    (a) How much do we have in the account after 8 years?

    (b) In how many years will the investment grow to $5000?

Module 8

Systems of Equations

8.1 Systems of Linear Equations

Here we consider problems that involve sets of linear equations involving two or more variables.

> A **system of linear equations** is a set of linear equations involving some number of variables. The number of equations need not equal the number of variables. A **solution** to a system is a set of values for the variables that satisfy *all* the equations in the system.

Why should we be interested in systems of linear equations? It turns out that many situations in life can be described by systems of equations of various sorts. Here is one example where systems of linear equations are used:

One of the primary functions of air traffic control is to make sure that airplanes don't crash into each other in the air. The path of each airplane is tracked and described by an algebraic equation. Then the equations are compared to see if there are any points at which they intersect. That is, the air traffic controller tries to find a solution for the system of equations that describe the routes of a set of airplanes—if there is one, the airplanes are on a collision course! The equations that arise may be linear (if a plane is flying in a straight line) or not (if a plane is circling the airport, for example).

Here are three examples of systems of two linear equations:

$2x + y = 10$ $x - y = 5$	The values $x = 5$ and $y = 0$ yield a solution for the system, since $2(5) + 0 = 10$ and $5 - 0 = 5$. The **unique solution** to the system is the *pair* of values $(x, y) = (5, 0)$.

| Not every system of equations has a solution. This system has no solution since any pair of values for x and y can satisfy at most one of these equations. In this case, we say that the system is **inconsistent**. | $2x + y = 10$ $2x + y = 20$ |

| $x + z = 4$ $y - z = 5$ | Another possibility is that a system may have **infinitely many solutions**. In general, if a system has more unknowns than equations, and if there is a solution, then there will be infinitely many solutions. In this system, for any value of z we pick, there are corresponding values for x and y satisfying the equations. Of course there are infinitely many choices for z, so there are infinitely many solutions to the system. |

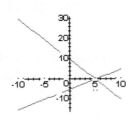

Systems of equations and their solutions also have a geometric interpretation. Each system above involving two equations and two variables describes a pair of lines in the xy plane. If there is a solution, it is the point at which the lines intersect. (The system is inconsistent when the lines are distinct and parallel.)

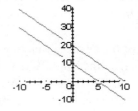

Before you go on to learn about solving systems of equations, take a brief tour of types of systems and their geometric interpretations. You will find the tour useful in understanding what types of solutions to expect and in interpreting your results.

8.1. SYSTEMS OF LINEAR EQUATIONS

Now is a great time to get an overview of the concepts of systems of equations and their solutions (or lack thereof). To do this, go through the slide show on the web.

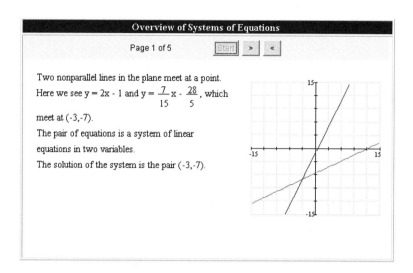

8.1. SYSTEMS OF LINEAR EQUATIONS

Additional Exercises

1. Which of the following pairs are solutions to the following system of equations?

$$2x - y = 4$$
$$x + 4y = 1$$

(a) $(-1, -6)$

(b) $(5, -1)$

(c) $(0, -4)$

(d) $\left(\frac{17}{9}, -\frac{2}{9}\right)$

2. Which of the following pairs are solutions to the following system of equations?

$$x - y = 0$$
$$x + y = 1$$

(a) $(-1, -1)$

(b) $(1, 0)$

(c) $(0, -2)$

(d) $\left(\frac{1}{2}, \frac{1}{2}\right)$

3. Consider the following system of equations:

$$2x^2 - y = 4$$
$$x^2 + 4y = 1$$

Is this a linear system of equations? Why? Replace x^2 with the variable t and write down the system of equations in terms of y and t. Is this system linear? Why or why not?

4. The equation of a circle of radius r and centered at (h, k) is $(x - h)^2 + (y - k)^2 = r^2$. Suppose the points $(0, 0)$, $(2, 0)$, and $(0, 2)$ lie on a circle. Then each point, when substituted into the equation of a circle, gives an equation. Substitute each of the three points to get the following system of equations:

$$h^2 + k^2 = r^2$$
$$4 - 4h + h^2 + k^2 = r^2$$
$$4 - 4k + h^2 + k^2 = r^2$$

Is this system linear in terms of the unknowns h, k, and r? Replace the expression $h^2 + k^2 - r^2$ with the variable t in each equation. Is the resulting system linear in h, k, and t?

8.2 Equivalent Systems

> Two systems of linear equations are said to be **equivalent** if they have the same solutions.

The following three systems are all equivalent to one another:

$$\begin{array}{lll} 4x + 5y = 13 & & \\ x - y = 1 & 2x + y = 5 & x + y = 3 \\ x + 2y = 4 & x + 2y = 4 & y = 1 \end{array}$$

From the third system of equations, we see that its second equation gives $y = 1$, and its first equation yields $x = 2$. We can verify by substitution that $(2, 1)$ is the solution to each of these systems.

Why are we interested in the concept of equivalent systems? One reason is that systems of equations are often presented in such a way that the solutions are not immediately apparent. To find the solutions, we might transform the system to another system for which the solutions are easier to find—while taking care not to add or delete solutions along the way! In the previous example, only the third system is presented in a way that makes it convenient to observe the solution set.

> Applying any combination of the following three steps to a system of linear equations produces an equivalent system of linear equations:
>
> 1. Multiply an equation by a non-zero constant.
>
> 2. Add one equation (or a non-zero multiple of it) to another.
>
> 3. Interchange two equations.

Here is an example of how the three steps can be applied.
Consider the system of three equations, E_1, E_2, E_3, in three unknowns:

$$\begin{array}{ll} E_1: & x + y = 2 \\ E_2: & 2x + 3y + z = 4 \\ E_3: & x + 2y + 2z = 6 \end{array}$$

- Replace E_2 with $E_2 - 2E_1$, and replace E_3 with $E_3 - E_1$. Leave E_1 as is. That uses item 2 from above twice. We now have a new equivalent system E_1, E_2, E_3:

$$\begin{aligned} E_1: &\quad x + y \phantom{{}+ 2z} = 2 \\ E_2: &\quad \phantom{x + {}} y + z = 0 \\ E_3: &\quad \phantom{x + {}} y + 2z = 4 \end{aligned}$$

- Replace E_3 with $E_3 - E_2$. Leave E_1 and E_2 as is. We have now produced the equivalent system:

$$\begin{aligned} E_1: &\quad x + y \phantom{{}+ z} = 2 \\ E_2: &\quad \phantom{x + {}} y + z = 0 \\ E_3: &\quad \phantom{x + y + {}} z = 4 \end{aligned}$$

From this last system we can deduce that $z = 4$, $y = -z = -4$, and $x = -y + 2 = -(-4) + 2 = 6$. Thus the unique solution to this system is $(x, y, z) = (6, -4, 4)$.

8.2. EQUIVALENT SYSTEMS

Additional Exercises

We shall refer to the following system in each question in this section. It will be useful to denote this system by S.

$$\begin{aligned} x - y + 2z &= 4 \\ x + 4y &= 1 \\ y - z &= 3 \end{aligned}$$

1. The following system of equations is equivalent to S.

$$\begin{aligned} x + 4y &= 1 \\ x - y + 2z &= 4 \\ y - z &= 3 \end{aligned}$$

   Explain why and how this system was derived from S.

2. Is the following system equivalent to S? Explain.

$$\begin{aligned} x + 4y &= 1 \\ -x + y - 2z &= -4 \\ y - z &= 3 \end{aligned}$$

3. Is the following system equivalent to S? Explain.

$$\begin{aligned} x + z &= 7 \\ x + 4y &= 1 \\ y - z &= 3 \end{aligned}$$

4. Is the following system equivalent to S? Explain.

$$\begin{aligned} -x + y - 2z &= -4 \\ x + 4y &= 1 \\ 2y - 2z &= 3 \end{aligned}$$

8.3 Infinitely Many Solutions

> A system of linear equations has **infinitely many solutions** when
>
> - There is at least one solution, and there are more unknowns than equations, or
>
> - There is at least one solution, and the system can be transformed to an equivalent one with more unknowns than equations.

Consider the following one-equation system:

$$3x - y = 12$$

There are many solutions to this equation; for example, $(x, y) = (4, 0)$ is a solution. Moreover, for any value we pick for one of the variables, there is a corresponding value for the other one so that together they form a solution. In fact, all the points on the line defined by this equation are solutions of the system. The solutions for this system may be described as the set of all pairs:

$$(x, y) = (t, 3t - 12), \text{ where } t \text{ is a real number.}$$

That is, for any value we choose for t, we get a pair that satisfies all the equations in the system. This way of expressing the solutions is called a **parametric form**, since it expresses all solutions in terms of the **parameter** t. For example, if $t = -1$, we get the solution $(-1, -15)$. If $t = 3$, we get the solution $(3, 3)$. You should verify that each of these particular solutions satisfies each equation in the system.

To see how we arrive at these parametric solutions when more variables are involved, consider this system of three equations in three unknowns:

$$\begin{aligned} x + y - z &= 1 \\ 2x - y &= 0 \\ 3y - 2z &= 2 \end{aligned}$$

If we add -2 times the first equation to the second equation and multiply the third equation by -1, we obtain the following equivalent system:

$$\begin{aligned} x + y - z &= 1 \\ -3y + 2z &= -2 \end{aligned}$$

This system has more equations than variables. Since the systems are equivalent, they have the same solution sets. Assigning a value to z determines the values of the other two variables. So

8.3. INFINITELY MANY SOLUTIONS

we will set z equal to some parameter, say t. From the second equation, we see that

$$-3y = -2 - 2z = -2 - 2t$$
$$y = \tfrac{2}{3} + \tfrac{2}{3}t$$

From the first equation, we now get this result:

$$x = 1 - y + z$$
$$x = 1 - (\tfrac{2}{3} + \tfrac{2}{3}t) + t$$
$$x = \tfrac{1}{3} + \tfrac{1}{3}t$$

Thus the solutions are all triples of the form $(x,y,z) = (\tfrac{1}{3} + \tfrac{1}{3}t, \tfrac{2}{3} + \tfrac{2}{3}t, t)$.

We have a geometric interpretation for this system also. Each of the equations describes a plane in three-dimensional space, and the solutions describe the straight line where these planes meet.

8.3. INFINITELY MANY SOLUTIONS

Additional Exercises

1. Find a parametric solution to the following system of equations by letting $z = t$ and solving for x and y in terms of t.

$$x - y + 2z = 4$$
$$y - z = 3$$

   Find one particular solution to the system by letting $t = -1$.

2. Find a parametric solution to the following system of equations by letting $y = s$ and solving for x and z in terms of s.

$$x - y + 2z = 4$$
$$y - z = 3$$

   What value of s would give the same particular solution found in the previous question?

3. Find a parametric solution to the following system of equations by letting $y = t$ and solving for x and z in terms of t.

$$x - y = 1$$
$$x + y + z = 0$$

   Find three particular solutions to the system by choosing three different values of t.

8.4 Solving Systems of Equations

In this section we describe two methods for finding solutions to a system of linear equations. These are known as the substitution method and the elimination method. In Module 9 we will study three other methods for solving systems of linear equations.

> The **substitution** method usually works well on systems of two equations in two unknowns. The essential idea here is that we solve one of the equations for one of the unknowns, and then substitute the result into the other equation.

Consider the following system:

$$2x + 3y = 5$$
$$x + y = 5$$

Solving the second equation for x gives $x = 5 - y$. Then substitute $5 - y$ for x into the first equation:

$$2(5 - y) + 3y = 5$$

Now solve for y:

$$10 - 2y + 3y = 5$$
$$10 + y = 5$$
$$y = -5$$

Going back to the solution for x in the previous equation, we now see that $x = 5 - (-5) = 10$. Thus the unique solution to this system is $(x, y) = (10, -5)$.

Note that it does not matter which equation we choose first and which we choose second. Just choose the most convenient one first! Also, it does not matter which unknown we choose first and which we choose second. Again, just choose the most convenient one first.

> The **elimination** method can be used for any number of equations in any number of unknowns. The idea is to transform the given system into an equivalent system from which the solutions are easily seen. We transform the system using the steps described in Section 8.2, Equivalent Systems.

Consider the system of three equations in three unknowns:

$$\begin{aligned} E_1: & \quad x + y \phantom{{}+z} = 2 \\ E_2: & \quad 2x + 3y + z = 4 \\ E_3: & \quad x + 2y + 2z = 6 \end{aligned}$$

- Replace E_2 with $E_2 - 2E_1$ and replace E_3 with $E_3 - E_1$. These steps eliminate x from the second and third equations. We now have a new equivalent system:

$$\begin{aligned} E_1: & \quad x + y \phantom{{}+2z} = 2 \\ E_2: & \quad \phantom{x+{}} y + z = 0 \\ E_3: & \quad \phantom{x+{}} y + 2z = 4 \end{aligned}$$

- Replace E_3 with $E_3 - E_2$. That eliminates y from the third equation and produces this system:

$$\begin{aligned} E_1: & \quad x + y \phantom{{}+z} = 2 \\ E_2: & \quad \phantom{x+{}} y + z = 0 \\ E_3: & \quad \phantom{x + y +{}} z = 4 \end{aligned}$$

From this last system, we can deduce that $z = 4$, $y = -z = -4$, and $x = -y + 2 = -(-4) + 2 = 6$. Thus, the unique solution to this system is $(x,y,z) = (6,-4,4)$.

Note some facts:

- We wrote each equation so that the variables appeared in the same order in each equation, and we also wrote the system of equations so that the variables lined up in a column. This makes the system easier to work with and prepares us for work with matrices in upcoming modules.

- Not all the systems are so nice and neat; after going through this process you might end up with fewer equations than variables, or with equations that contradict each other. In the first case, we have a system with infinitely many solutions; in the second, the system is inconsistent and so has no solutions at all. See Section 8.1, Systems of Linear Equations, and the Frequently Asked Questions on the web for a discussion of these situations and the solutions that arise through them.

_____

It's time to get to work! For each system that you are given in these applets, first find the solution(s) using one of the methods you learned in previous sections. Then determine whether there is no solution, a unique solution, or infinitely many solutions. Finally, enter your results in the applets to check them.

8.4. SOLVING SYSTEMS OF EQUATIONS

In the following applets you will receive systems of equations in two unknowns and in three unknowns. In each of them,

1. Pick a method to work through on paper.

2. Decide whether the system has a solution. If it has a solution, determine whether it is unique or there are infinitely many solutions. Click on the appropriate button.

3. Now follow the steps in the window that come up for each situation.

You may have questions about the results of your computations; if so, go to the FAQ page to get help.

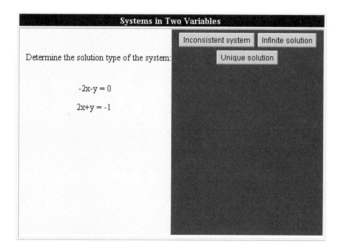

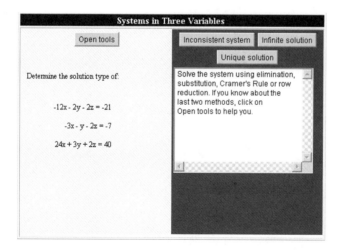

Additional Exercises

1. Solve the following system of equations using the substitution method.

$$x + 4y = 1$$
$$x - y = 4$$

Solve using the elimination method. Which method do you prefer for this system?

2. Solve the following system of equations using the substitution method.

$$3x + 4y = 1$$
$$2x - 5y = 0$$

Solve using the elimination method. Which method do you prefer for this system?

3. Solve the following system of equations.

$$x + y = 1$$
$$x - y - z = -1$$
$$3x + 3y + z = 1$$

4. Solve the following system of equations.

$$5x - y - 3z = -1$$
$$x - 2y - 2z = -1$$
$$9x - 3y - 5z = -5$$

5. Determine the equation of the circle that contains the points (0,0), (2,0), and (0,2). What is the radius of this circle? (See Exercise 4 in Section 8.1.)

6. A teacher is about to take her class to the zoo. She can purchase either 25 individual tickets at $7 each or get a group rate, which is $36 plus $5.50 per ticket.

   (a) Set up a system of two equations describing this situation.
   (b) Solve the system.
   (c) What does the solution to the system represent in this problem?

8.4. SOLVING SYSTEMS OF EQUATIONS

7. Betsy has lots of stuffed animals and several baskets in which to store them. If she puts three animals in each basket, she has one animal left over. If she puts four in each basket, she winds up with two empty baskets. How many animals and baskets does Betsy have?

8. A group of friends is planning a one-day driving trip, so they are pricing rental vehicles. Drive Away Cars offers vans for $30 per day plus 20¢ per mile. Hit the Road Rentals has vans for $25 per day plus 25¢ per mile. Solve a system of equations to determine the mileage ranges for which each company is less expensive.

9. Harold has $3 in change consisting of nickels, dimes, and quarters, with 24 coins in all. The total number of nickels and dimes is twice the number of quarters. How many of each type of coin does Harold have?

10. Kevin has change consisting of nickels, dimes, and quarters, with 24 coins in all. The total number of nickels and dimes is twice the number of quarters. What is the least amount of money Kevin could have? What is the most?

8.5 Mixture Problems

Problems that involve mixing together components such as solutions or alloys to produce a new mixture satisfying certain conditions often lead to a system of linear equations. Those problems that involve only two components usually lead to a system of two equations that can be solved by a variety of methods.

Two silver alloys containing 40% and 70% pure silver by mass, respectively, are to be mixed to form a new alloy. How many grams of each are required to produce 20 g of a new alloy that is 55% silver?	
Using x and y to represent the grams of each alloy to be used, we can form these two equations. The first represents the amounts of pure silver being combined to produce $(0.55)(20) = 11$ g of silver in the new alloy. The second equation simply represents the total mass. We can also clear the first equation of the decimal point by multiplying both sides by 10.	$0.4x + 0.7y = 11$ $x + y = 20$ or $4x + 7y = 110$ $x + y = 20$
Any of the methods discussed in this module could be applied. Here we use elimination since it is easy to scale the second equation to eliminate either x or y. We multiply the second equation by -4 to eliminate the variable x.	$\begin{aligned} 4x + 7y &= 110 \\ -4x - 4y &= -80 \\ \hline 3y &= 30 \end{aligned}$
At this point we know that $y = 10$ g and, since the total mass must be 20 g, $x = 10$ g.	Solution: Use 10 g of the 40% alloy and 10 g of the 70% alloy.

Go to the web to try some mixture problems. All problems in the exercise require you to develop and solve a system of two linear equations in two variables. Help is provided if you have difficulty determining the equations. A complete solution is provided for you to check your work.

The following exercise presents similar mixture problems. A Help window is available to explain the two equations generated. A Solve window shows how that system of two equations can be solved by elimination.

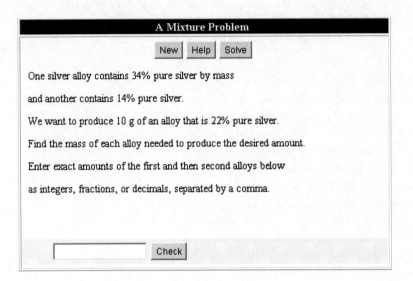

8.5. MIXTURE PROBLEMS

Additional Exercises

1. Two gold alloys containing 20% and 60% pure gold by mass are to be mixed to form a new alloy. How many grams of each are required to produce 10 g of a new alloy that is 50% gold?

2. Vanilla extract is 25% alcohol, and rice cooking wine is 16%. A recipe calls for 4 oz of a mixture that is 20% alcohol. Find the amounts needed to produce such a mixture.

3. An herb distributor wishes to create a seasoning mix of oregano, basil, and rosemary. She can purchase oregano at 50¢ per ounce, basil at 70¢ per ounce, and rosemary at 90¢ per ounce. How much of each herb is needed to produce 3 ounces of seasoning mix that will cost her 60¢ per ounce if the seasoning mix is to contain twice as much oregano as basil?

8.6 Parabola from Three Points

How many parabolas can pass through three given points in the plane? If we allow parabolas that can be tilted to open in any desired direction, then, assuming the three points are not collinear, there are infinitely many. However, if we restrict our attention to parabolas that open up or down, then there is only one parabola passing through three given noncollinear points. The reasons for using three points lie in the form of the formula that describes parabolas that open up or down: $y = ax^2 + bx + c$. If we are trying to determine the appropriate values of a, b, and c, then each time we replace x and y in this formula with the coordinates of a given point, we get a linear equation in three unknowns: the coefficients a, b, and c. With three unknowns, we need three points to generate three linear equations. Fewer equations could not produce a unique solution. The resulting system of three equations will have a solution as long as the points are not collinear. Here is an example:

$y = ax^2 + bx + c$	
Given Point	**Resulting Equation**
$(3, -1)$	$9a + 3b + c = -1$
$(2, 4)$	$4a + 2b + c = 4$
$(-1, 5)$	$a - b + c = 5$

On the web is an exercise to determine the equation of a parabola that passes through three given points. The technique for solving each problem is identical to that in the example above, and you should use that example as a guide.

In this next exercise you must find the parabola through the three points given. Help and solutions are available. The system of linear equations can be solved by any method you choose. If you want to use matrix methods from Module 9, the pop-up matrix tool is included.

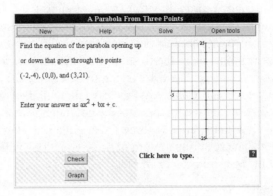

8.6. PARABOLA FROM THREE POINTS

Additional Exercises

1. Find the equation of a parabola that passes through the points $(-2, 12)$, $(0, 2)$, and $(3, 7)$.

2. Find the equation of a parabola that passes through the points $(-3, 20)$, $(1, -4)$, and $(4, -1)$.

3. Find the equations of all parabolas that pass through the points $(1, 1)$ and $(-1, 1)$. (Hint: Obtain a system of two linear equations in three variables, and find a parametric solution to the system.)

Module 9

Matrices

9.1 Matrices

> A **matrix** is a rectangular array of entries. The entries in the matrices we will use will all be real numbers.

Matrices come in all possible rectangular shapes. Following are a number of examples:

$$\begin{bmatrix} 6 & -3 & 18 & 1 \end{bmatrix} \quad \begin{bmatrix} -5 \\ 5 \\ 9 \end{bmatrix} \quad \begin{bmatrix} 17 & 0 & -5 \\ 11 & 2 & 3 \end{bmatrix} \quad \begin{bmatrix} 3 & 2 \\ 9 & 8 \\ 0 & 1 \\ 6 & 7 \end{bmatrix}$$

In general, we denote a matrix as follows:

$$\begin{bmatrix} a_{11} & a_{12} & \cdots & a_{1n} \\ a_{21} & a_{22} & \cdots & a_{2n} \\ \vdots & \vdots & \ddots & \vdots \\ a_{m1} & a_{m2} & \cdots & a_{mn} \end{bmatrix}$$

The entries of the matrix are organized in horizontal rows and vertical columns. We generally use the symbol a_{ij} to represent the entry (or element) of the matrix A in row i and column j.

> A matrix with m rows and n columns has **dimensions $m \times n$** and is read "m by n."

The four matrices above are of dimensions 1×4, 3×1, 2×3, and 4×2, respectively.

> A **square matrix** is one with the same number of rows and columns. If a square matrix has n rows and n columns, we say that it has **order n.** Square matrices also have a special set of entries: those on the diagonal from top left to bottom right. This is called the **principal** (or **main**) **diagonal**, and its elements are called the **principal** (or **main**) **diagonal entries**.

Example: The matrix on the right is a square matrix of order 3, and its main diagonal entries are 1, −1, and 5.
$$\begin{bmatrix} 1 & 2 & 0 \\ 0 & -1 & 3 \\ 2 & 4 & 5 \end{bmatrix}$$

Algebra of Matrices

Arithmetic operations are defined for matrices of appropriate dimensions. We describe the requirements and the operations as follows:

> **Scalar multiplication:** To multiply a matrix A by a number c, multiply each entry of A by the number c. The result is a matrix of the same dimensions as A.

Example:

$$5 \begin{bmatrix} 2 & -4 & 1 \\ 3 & 0 & -2 \end{bmatrix} = \begin{bmatrix} 5(2) & 5(-4) & 5(1) \\ 5(3) & 5(0) & 5(-2) \end{bmatrix} = \begin{bmatrix} 10 & -20 & 5 \\ 15 & 0 & -10 \end{bmatrix}$$

> **Addition (and subtraction):** If two matrices have the same dimensions, we may add them; we do so by adding elements in corresponding positions. In this case the result is a matrix with the same dimensions as those of the original matrices.

Example:

$$\begin{bmatrix} 6 & -8 \\ 11 & 5 \end{bmatrix} + \begin{bmatrix} -3 & 1 \\ 0 & 12 \end{bmatrix} = \begin{bmatrix} 6-3 & -8+1 \\ 11+0 & 5+12 \end{bmatrix} = \begin{bmatrix} 3 & -7 \\ 11 & 17 \end{bmatrix}$$

Matrix subtraction is not actually a new operation (just as subtraction of real numbers is not a separate operation). To subtract matrix B from matrix A, simply multiply matrix B by -1 (using the scalar multiplication described above) and then add them:

$$A - B = A + (-1)B$$

For example:

$$\begin{bmatrix} 5 & 9 \\ -3 & 4 \\ 8 & -6 \end{bmatrix} - \begin{bmatrix} 2 & 0 \\ 7 & -5 \\ 1 & -1 \end{bmatrix} = \begin{bmatrix} 5-2 & 9-0 \\ -3-7 & 4+5 \\ 8-1 & -6+1 \end{bmatrix} = \begin{bmatrix} 3 & 9 \\ -10 & 9 \\ 7 & -5 \end{bmatrix}$$

9.1. MATRICES

> **Matrix multiplication:** The product AB of two matrices A and B is defined if A has as many columns as B has rows. For an $m \times n$ matrix A and an $n \times p$ matrix B, the product AB will have dimensions $m \times p$. The entry in the ith row and jth column of AB is $a_{i1}b_{1j} + a_{i2}b_{2j} + \cdots + a_{in}b_{nj}$. This sum is the result of multiplying each entry in the ith row of A by its corresponding entry in the jth column of B and then adding those n products together.

The order of the factors is critical here. For example, we cannot multiply a 3×4 matrix by a 2×3 matrix in that order. However, we can multiply a 2×3 matrix by a 3×4 matrix, as the next activity demonstrates. Note how each of the eight entries in the 2×4 product is calculated.

The definition of matrix multiplication and understanding how to compute an entry in the product is a bit daunting at first. The following computer demonstration should help to make it clearer.

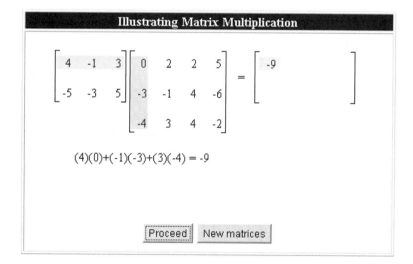

You can practice matrix multiplication in the next computer activity. First determine whether two matrices can be multiplied together, and then, if so, find each entry of the product.

Now practice multiplying matrices. If the multiplication *can* be carried out, find the dimensions of the resulting product matrix, enter the number of rows and columns and click Can , and continue by entering the entries for the product. If you have a problem, click

Reveal and you will get guidance. If you determine that the product *cannot* be carried out, click Can't.

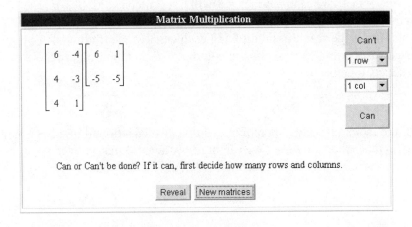

Applications to Systems of Linear Equations

A major application of matrices is to the solution of systems of linear equations. The first order of business in this connection is to define appropriate matrices related to a system.

Consider the following system:

$$\begin{aligned} 2x - 3y + z &= -1 \\ 3x + y + 2z &= 0 \\ x - 5y &= 2 \end{aligned}$$

There are two important matrices related to this system: The **coefficient matrix** is composed of the coefficients of the variables. Each column lists the coefficients of one of the variables, and each row lists the coefficients in one of the equations. In this case, we get a square matrix since there are three equations in three unknowns. There could, of course, be any size matrix here—a system of m equations in n unknowns will yield a matrix of dimension $m \times n$.

$$\begin{bmatrix} 2 & -3 & 1 \\ 3 & 1 & 2 \\ 1 & -5 & 0 \end{bmatrix}$$

We also have the **augmented matrix**, which is the coefficient matrix with an additional column "glued on" to the right; this column consists of the numbers to the right of the equal signs in the equations.

$$\begin{bmatrix} 2 & -3 & 1 & -1 \\ 3 & 1 & 2 & 0 \\ 1 & -5 & 0 & 2 \end{bmatrix}$$

To see how to use matrices to solve systems, go to Section 9.2, Systems of Equations and Matrices.

9.1. MATRICES

Additional Exercises

1. Assume the following:

$$A = \begin{bmatrix} 0 & -1 \\ 2 & 1 \end{bmatrix} \quad B = \begin{bmatrix} -2 & 1 \\ 0 & 1 \end{bmatrix} \quad C = \begin{bmatrix} 3 & 2 \\ -5 & 0 \end{bmatrix}$$

Compute, whenever possible, the following. For those that you cannot compute, explain why not.

(a) $A + B$

(b) $C - B$

(c) AC

(d) $A(B + C)$

(e) $3A - 2B$

2. Assume the following:

$$A = \begin{bmatrix} 0 \\ 2 \end{bmatrix} \quad B = \begin{bmatrix} -2 \\ 0 \end{bmatrix} \quad C = \begin{bmatrix} 1 & -2 \\ 4 & 1 \end{bmatrix}$$

Compute, whenever possible, the following. For those that you cannot compute, explain why not.

(a) $A + B$

(b) $C - B$

(c) AC

(d) $(A + B)C$

(e) $3A - 2B$

INTERACTIVE COLLEGE ALGEBRA: A WEB-BASED COURSE © 2005 Key College Publishing

3. Assume the following:

$$A = \begin{bmatrix} 0 & 2 \\ 4 & -2 \\ 1 & 3 \end{bmatrix} \quad B = \begin{bmatrix} -2 & 0 & 1 \\ 0 & 4 & 3 \\ -1 & -1 & 2 \end{bmatrix} \quad C = \begin{bmatrix} 1 & -2 \\ 4 & 1 \\ 0 & -2 \end{bmatrix}$$

Compute, whenever possible, the following. For those that you cannot compute, explain why not.

(a) $A + C$

(b) $C - B$

(c) BC

(d) $(A + C)B$

(e) $3A - 2B$

4. Write the coefficient matrix and the augmented matrix for the following system of equations:

$$\begin{aligned} 2x - y &= 4 \\ x + 4y &= 1 \end{aligned}$$

9.1. MATRICES

5. Write the coefficient matrix and the augmented matrix for the following system of equations:

$$\begin{aligned} 4x - y + 3z &= 4 \\ x + 2y - 5z &= 1 \\ 2x - 7y &= 8 \end{aligned}$$

6. If A is an $m \times n$ matrix and B is a $p \times q$ matrix, under what conditions do both the products AB and BA exist? Under these conditions, what are the sizes of AB and BA?

7. For a system of k linear equations in r variables, what are the dimensions of the coefficient matrix and the augmented matrix?

9.2 Systems of Equations and Matrices

We have seen how to solve certain systems of equations using the substitution method and the elimination method. However, these are not the most efficient methods in many cases, especially if the system is large. Other, often better, methods use matrices. Here we will describe how to solve systems of linear equations using matrices and employing a technique known as **row-reduction**.

When we form the augmented matrix of a system of linear equations, we are essentially recording the coefficients of the variables and constant terms. In solving a linear system using the elimination method, we are allowed to interchange any two equations, add a multiple of one equation to another, or multiply every term of an equation by a non-zero number. These equation changes correspond to the following **row operations** on the augmented matrix:

Interchange two rows	This operation is equivalent to changing the order in which we write two equations of the system.
Multiply (or divide) a row by a non-zero constant	This operation simply multiplies an equation by a constant; it does not affect the solution.
Add a multiple of one row to another (This operation, of course, allows you to subtract one row from another, because it is the same as adding -1 times the row.)	This operation is exactly what we do in the elimination method. This replacement of an existing equation with itself plus a multiple of another equation does not change the solution of the system.

It is convenient to introduce some shorthand notation to describe the row operation that has been performed. You may find it useful to use this notation when working on problems. It will help you to remember what step you did at each stage of the process in row-reducing a matrix.

Interchange row i with row j	$R_i \longleftrightarrow R_j$
Multiply row i by the non-zero number c	cR_i
Add c times row j to row i	$R_i + cR_j$

The idea behind the elimination method is to produce a system of simpler equations in which most of the variable coefficients are zero and then read off the solution to the system, either directly or by backward substitution. For example, with a system of three equations in three unknowns, we will try to produce a system having an equation such as $z = -2$. When we translate this to row operations on the augmented matrix, we want to produce a matrix in which there are many zeros and be able to read off the solutions easily. This can be done in many ways, but we'd like to have a method that is easily described. To this end, we shall perform row operations on the augmented matrix to row-reduce it to **row-echelon form**.

> A matrix is in row-echelon form if all of the following are true:
>
> - The first non-zero entry in any row is 1. This is known as a **leading term**.
>
> - All entries in a column below a leading term are 0.
>
> - If a row has a leading term, then the leading term of any subsequent row must appear to the right.
>
> - Any rows containing only zeros appear as the bottom rows of the matrix.

Note that although we will usually want to transform an augmented matrix into row-echelon form to solve a system of equations, the process can be applied abstractly to any matrix.

You should be able to spot if a matrix is in row-echelon form by inspection. Try the exercise on the web page. Keep in mind the definition of row-echelon form of a matrix.

The next exercise will give you practice in determining whether a matrix is in row-echelon form. The answer is a simple yes or no. Try plenty of questions in this exercise.

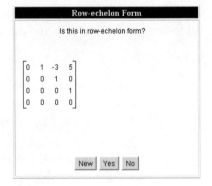

9.2. SYSTEMS OF EQUATIONS AND MATRICES

Let's summarize the technique of solving a system of linear equations by row-reducing the augmented matrix.

> 1. Write the augmented matrix of the system.
> 2. Row-reduce the augmented matrix to row-echelon form.
> 3. Write the new equivalent system that is defined by the matrix in row-echelon form.
> 4. The solution (or lack thereof) is now apparent.

Example: Solve the following system of equations:
$$\begin{aligned} 3x + 2y - z &= 1 \\ x + 2y + 2z &= 0 \\ 2x + y - 3z &= -1 \end{aligned}$$

Step 1: Write the augmented matrix of the system:
$$\begin{bmatrix} 3 & 2 & -1 & 1 \\ 1 & 2 & 2 & 0 \\ 2 & 1 & -3 & -1 \end{bmatrix}$$

Step 2: Row-reduce the augmented matrix:

$$\begin{bmatrix} 1 & 2 & 2 & 0 \\ 3 & 2 & -1 & 1 \\ 2 & 1 & -3 & -1 \end{bmatrix} \quad R_1 \longleftrightarrow R_2$$

$$\begin{bmatrix} 1 & 2 & 2 & 0 \\ 0 & -4 & -7 & 1 \\ 0 & -3 & -7 & -1 \end{bmatrix} \quad R_2 - 3R_1,\ R_3 - 2R_1$$

$$\begin{bmatrix} 1 & 2 & 2 & 0 \\ 0 & 1 & \frac{7}{4} & -\frac{1}{4} \\ 0 & 1 & \frac{7}{3} & \frac{1}{3} \end{bmatrix} \quad R_2/(-4),\ R_3/(-3)$$

$$\begin{bmatrix} 1 & 2 & 2 & 0 \\ 0 & 1 & \frac{7}{4} & -\frac{1}{4} \\ 0 & 0 & \frac{7}{12} & \frac{7}{12} \end{bmatrix} \quad R_3 - R_2$$

$$\begin{bmatrix} 1 & 2 & 2 & 0 \\ 0 & 1 & \frac{7}{4} & -\frac{1}{4} \\ 0 & 0 & 1 & 1 \end{bmatrix} \quad \frac{12R_3}{7}$$

Step 3: Rewrite the system using the row-reduced matrix:

$$x + 2y + 2z = 0$$
$$y + \tfrac{7}{4}z = -\tfrac{1}{4}$$
$$z = 1$$

Step 4: The solution is found by going from the bottom equation up:

$$z = 1$$
$$y = -\tfrac{7}{4}z - \tfrac{1}{4} = -2$$
$$x = -2y - 2z = 2$$

So, the unique solution is $(x, y, z) = (2, -2, 1)$.

One problem you will encounter is deciding which of these row operations, and in which order, you should apply to a matrix to row-reduce it to row-echelon form. Although algorithms do exist for transforming a matrix to row-echelon form, in our examples and problems it is easier to determine which operation to perform by observation. This, of course, takes practice, and the more experience you gain the easier it will become.

There is a demonstration on the web page at this point showing a matrix being reduced to row-echelon form. Follow along, observing what row operation is performed at each stage. Would you have made the same choice?

This demonstration gives many examples of row-reducing a matrix to row-echelon form. Work through several examples to make sure you feel comfortable with the process. Click on [Step] to see which operation will be performed, and then click on [Do it] to carry out the operation.

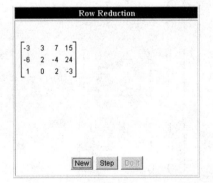

9.2. SYSTEMS OF EQUATIONS AND MATRICES

Try out the following demonstration to make sure you understand how to obtain the solutions of a system of linear equations from the row-echelon form of the augmented matrix. Each example shows a row-echelon form of the augmented matrix of a system of three linear equations in three unknowns, x, y, and z.

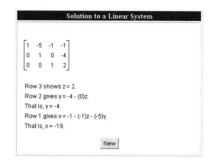

Now go back to Module 8 on solving systems of equations using the row-reduction method.

Additional Exercises

1. Solve the following system of linear equations by finding the augmented matrix and row-reducing to row-echelon form.
$$x + 4y = 1$$
$$x - y = 4$$

2. Solve the following system of linear equations.
$$3x + 4y = 1$$
$$2x - 5y = 0$$

3. Solve the following system of linear equations.

$$5x - y - 3z = -1$$
$$x - 2y - 2z = -1$$
$$9x - 3y - 5z = -5$$

4. Determine a parametric solution to the following system of equations using row-reduction of the augmented matrix of the system.

$$3x - y - 3z = -1$$
$$2x - 3y - 3z = -5$$

9.3 Determinants

Associated with each square matrix is a number whose value is determined by the entries of the matrix. This number is called the **determinant** of the matrix. If A is a square matrix, we denote its determinant by $|A|$.

The determinant of an arbitrary-sized square matrix will not be discussed in this course. However, we shall make use of the determinant of 2×2 and 3×3 matrices. We now give a method for computing the value of these determinants.

To find the determinant of a 2×2 matrix, subtract the product of the entries on the top right and bottom left from the product of the entries on the main diagonal.	$\begin{array}{cc} + & - \\ \begin{vmatrix} -2 & -9 \\ 4 & -6 \end{vmatrix} \end{array}$ $(-2)(-6) - (4)(-9)$ $= 48$

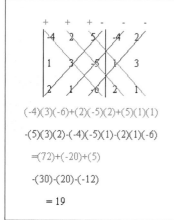 $(-4)(3)(-6)+(2)(-5)(2)+(5)(1)(1)$ $-(5)(3)(2)-(-4)(-5)(1)-(2)(1)(-6)$ $=(72)+(-20)+(5)$ $-(30)-(20)-(-12)$ $= 19$	Finding the determinant of a 3×3 matrix involves six products, each consisting of three factors. Three of the products are added in, and three are subtracted. One mechanism for doing the calculation is the method shown here. Copy the first two columns to the right of the array and form the six products (on the diagonals) as shown. The first three are added, and each of the last three is subtracted.

The exercises for computing determinants open up in new windows when you click on the buttons $\boxed{2 \times 2}$ *or* $\boxed{3 \times 3}$. *First try the 2 × 2 determinants. On paper, evaluate the determinant, then check your answer with that given by the computer.*

If you want to practice some determinant calculations, select one of these sizes: $\boxed{2 \times 2}$ $\boxed{3 \times 3}$

Additional Exercises

1. Compute the determinant $|A|$ for each of the following:

    (a) $A = \begin{bmatrix} 0 & -1 \\ 2 & 1 \end{bmatrix}$ $|A| =$

    (b) $A = \begin{bmatrix} -2 & 1 \\ 0 & 1 \end{bmatrix}$ $|A| =$

    (c) $A = \begin{bmatrix} 3 & 2 \\ -5 & -9 \end{bmatrix}$ $|A| =$

2. Compute the determinant:

$$\begin{vmatrix} -2 & 0 & 1 \\ 0 & 4 & 3 \\ -1 & -1 & 2 \end{vmatrix}$$

3. Compute the determinant:

$$\begin{vmatrix} 1 & -2 & 0 \\ 4 & 1 & -1 \\ 0 & -2 & 0 \end{vmatrix}$$

4. The matrix B is obtained from matrix A by interchanging rows 1 and 2 of A.

$$B = \begin{bmatrix} 0 & -2 & 0 \\ 4 & 1 & -1 \\ 1 & -2 & 0 \end{bmatrix}$$

$$A = \begin{bmatrix} 4 & 1 & -1 \\ 0 & -2 & 0 \\ 1 & -2 & 0 \end{bmatrix}$$

Compute $|A|$ and $|B|$. What is the connection between $|A|$ and $|B|$?

5. The matrix B is obtained from matrix A by multiplying row 1 of A by 3.

$$B = \begin{bmatrix} 12 & 3 & -3 \\ 0 & -2 & 0 \\ 1 & -2 & 0 \end{bmatrix}$$

$$A = \begin{bmatrix} 4 & 1 & -1 \\ 0 & -2 & 0 \\ 1 & -2 & 0 \end{bmatrix}$$

Compute $|A|$ and $|B|$. What is the connection between $|A|$ and $|B|$?

9.3. DETERMINANTS

6. The matrix B is obtained from matrix A by adding to row 1 of A 3 times row 3.

$$B = \begin{bmatrix} 7 & -5 & -1 \\ 0 & -2 & 0 \\ 1 & -2 & 0 \end{bmatrix}$$

$$A = \begin{bmatrix} 4 & 1 & -1 \\ 0 & -2 & 0 \\ 1 & -2 & 0 \end{bmatrix}$$

Compute $|A|$ and $|B|$. What is the connection between $|A|$ and $|B|$?

7. Let matrix B be obtained from matrix A by some elementary row operation. Determine the connection between $|A|$ and $|B|$ for each type of row operation.

9.4 Cramer's Rule

In this section we will describe a method of solving systems of linear equations using determinants. This method is known as **Cramer's Rule**.

In order to use Cramer's Rule, two prerequisites must be met:

- The system must have the same number of equations as variables, that is, the coefficient matrix of the system must be square.

- The determinant of the coefficient matrix must be non-zero. The reason for this will become apparent as we describe the method.

The steps for applying Cramer's Rule are as follows:

---

1. Write the coefficient matrix of the system A and find its determinant $|A|$.

2. Suppose the first variable of the system is x. Then write the matrix A_x as follows: Substitute the column of numbers to the right of the equal signs instead of the first column of A. Now compute the determinant $|A_x|$.

3. The value of x in the solution is $|A_x|/|A|$.

4. Repeat for the remaining variables. In each case, substitute the column of numbers for the column of A that corresponds to the variable you are finding.

If the variables are x, y, and z, then the solutions will be

$$(x, y, z) = \left(\frac{|A_x|}{|A|}, \frac{|A_y|}{|A|}, \frac{|A_z|}{|A|} \right)$$

We will demonstrate Cramer's Rule with the following system:

$$\begin{aligned} x + 2y + 3z &= 1 \\ -x + 2z &= 2 \\ -2y + z &= -2 \end{aligned}$$

Step 1: The coefficient matrix of this system is $A = \begin{bmatrix} 1 & 2 & 3 \\ -1 & 0 & 2 \\ 0 & -2 & 1 \end{bmatrix}$ and $|A| = 12$.

Step 2: $A_x = \begin{bmatrix} 1 & 2 & 3 \\ 2 & 0 & 2 \\ -2 & -2 & 1 \end{bmatrix}$ and $|A_x| = -20$.

Step 3: Therefore, $x = |A_x|/|A| = -20/12 = -5/3$.

Step 4: Using the same method, the values for the remaining two variables, y and z, are computed as follows:

$A_y = \begin{bmatrix} 1 & 1 & 3 \\ -1 & 2 & 2 \\ 0 & -2 & 1 \end{bmatrix}$ and $y = |A_y|/|A| = 13/12$.

$A_z = \begin{bmatrix} 1 & 2 & 1 \\ -1 & 0 & 2 \\ 0 & -2 & -2 \end{bmatrix}$ and $z = |A_z|/|A| = 2/12 = 1/6$.

The solutions are $(x, y, z) = (-5/3, 13/12, 1/6)$.

Click on the link practice below to jump back to the module on solving systems of equations. There, practice solutions of systems of equations using Cramer's Rule.

Now practice solving the systems of equations in Section 8.4 using Cramer's Rule.

Additional Exercises

1. Use Cramer's Rule to solve the following system of linear equations.

$$\begin{aligned} x + 4y &= 1 \\ x - y &= 4 \end{aligned}$$

2. Solve the following system of linear equations using Cramer's Rule.

$$\begin{aligned} 3x + 4y &= 1 \\ 2x - 5y &= 0 \end{aligned}$$

3. Solve the following system of linear equations using Cramer's Rule.

$$\begin{aligned} 5x - y - 3z &= -1 \\ x - 2y - 2z &= -1 \\ 9x - 3y - 5z &= -5 \end{aligned}$$

9.5 Inverse Matrices

The $n \times n$ matrix for which every diagonal entry is 1 and every non-diagonal entry is 0 is called the **$n \times n$ identity matrix** and is denoted by I_n. Its role in matrix multiplication is analogous to the role of the number 1 in multiplication of real numbers. That is, for every $n \times n$ matrix A,

$$AI_n = I_n A = A$$

The identity matrix allows us to define the concept of an inverse matrix of a square matrix.

> Suppose A and B are $n \times n$ matrices. If $AB = BA = I_n$, then we say that A is an **inverse matrix** to B and B is an inverse matrix to A.

If an inverse matrix exists, then it is unique. That is, we *cannot* have different matrices B and C both having the property that $AB = BA = I_n$ and $AC = CA = I_n$. We also note that not every square matrix has an inverse, for example:

$$\begin{bmatrix} 1 & 1 \\ 0 & 0 \end{bmatrix}$$

This leads to two questions: Under what conditions does a matrix have an inverse? and, If there is an inverse matrix, how do we find it? The first question has an easy answer (but the reason is beyond the scope of this course).

> An $n \times n$ matrix A has an inverse only if the determinant of A is not zero.

Several methods are used to find the inverse of a matrix. For a 2×2 matrix, we can find the inverse matrix as follows:

- Switch the diagonal entries.
- Negate the non-diagonal entries.
- Divide each entry by the determinant of the matrix.

Example: Consider the following matrix:

$$\begin{bmatrix} 1 & 2 \\ 3 & 5 \end{bmatrix}$$

The determinant is -1, so this matrix does have an inverse. Let's apply the method to find the inverse. First, switch the diagonal entries to get

$$\begin{bmatrix} 5 & 2 \\ 3 & 1 \end{bmatrix}$$

Second, negate the other entries to get

$$\begin{bmatrix} 5 & -2 \\ -3 & 1 \end{bmatrix}$$

Finally, divide each entry by the determinant -1:

$$\begin{bmatrix} -5 & 2 \\ 3 & -1 \end{bmatrix}$$

The method we have described works *only* for 2×2 matrices. Next we'll describe a method that works for any square matrix. For an $n \times n$ matrix, we can use row-reduction techniques to obtain the inverse matrix. Here are the steps to find the inverse of a matrix:

> 1. Given an $n \times n$ matrix A, write the $n \times n$ identity matrix to the right of A and denote this by $A \mid I_n$.
>
> 2. Perform row operations on the rows of the new matrix $A \mid I_n$ with the goal of attaining I_n on the left instead of A.
>
> 3. The inverse of A is the resulting matrix on the right.
>
> 4. Check that your new matrix is indeed the inverse of A by multiplying them together to get I_n.

With this method of computing the inverse, it turns out that we do not actually have to find the determinant of A to decide whether it has an inverse. We can go through the procedure of finding the inverse, and if the procedure actually is able to produce an identity matrix on the left, then we know that the matrix on the right is indeed the inverse of the original matrix.

9.5. INVERSE MATRICES

Next on the web page is a demonstration of an attempt to find an inverse matrix using row-reduction. Work through some examples, and make sure you understand the process. Note that we never compute the determinant of the given matrix to see if it has an inverse, since that will be determined by the method.

We demonstrate how to find the inverse matrix of a 3×3 matrix using row-reduction. Click on Step to see what elementary row operation will be carried out, and then click Do it to perform the operation. Make sure that you understand every step. You can also click Auto to have the applet go through each step, pausing between stating what row operation is to be done and carrying out the operation. If this method is too slow, click Stop. At that point, you can go back to the manual method to continue finding the inverse.

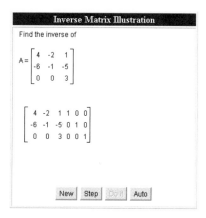

You can practice finding inverse matrices using the next exercise. You will be given a matrix and asked to find its inverse. You should work out your answer on paper, or use the matrix manipulative provided in the tools. Compare your answer with the computer's.

You should now try to find the inverse of a matrix using this method. In this next exercise, you will be given a matrix. On paper, or using the matrix manipulator from the tools provided, use row-reduction to obtain the inverse matrix. Once you have done that, click on Show me to see the inverse and compare your answer.

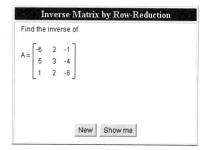

Additional Exercises

1. Find, if possible, A^{-1}, B^{-1}, and C^{-1} where

$$A = \begin{bmatrix} 0 & -1 \\ 2 & 1 \end{bmatrix} \quad B = \begin{bmatrix} -2 & 1 \\ 0 & 1 \end{bmatrix} \quad C = \begin{bmatrix} 3 & 2 \\ -5 & 0 \end{bmatrix}$$

2. Find A^{-1}, if possible, where

(a) $A = \begin{bmatrix} 0 & 1 & 1 \\ 1 & 0 & 0 \\ 0 & 1 & 0 \end{bmatrix}$

(b) $A = \begin{bmatrix} 0 & 0 & 1 \\ 0 & 1 & 3 \\ 1 & -1 & 2 \end{bmatrix}$

(c) $A = \begin{bmatrix} -2 & 0 & 1 \\ 0 & 4 & 3 \\ -1 & -1 & 2 \end{bmatrix}$

(d) $A = \begin{bmatrix} 1 & -3 & 3 \\ 2 & 1 & 0 \\ 3 & -2 & 3 \end{bmatrix}$

9.6 Systems of Equations and Inverse Matrices

Suppose you are given an equation in one variable, such as $4x = 9$. Then you will find the value of x that solves this equation by multiplying the equation by the multiplicative inverse of 4, which is $1/4$:

$$\left(\tfrac{1}{4}\right)4x = \left(\tfrac{1}{4}\right)9$$
$$x = \tfrac{9}{4}$$

Sometimes we can do something very similar to solve systems of linear equations; in this case, we will use the **inverse of the coefficient matrix**. In order to solve a system by this method, two conditions must be met:

- The system must have the same number of equations as variables; that is, the coefficient matrix of the system must be square.

- The determinant of the coefficient matrix must be non-zero; this is because the inverse of a square matrix exists precisely when its determinant is non-zero.

The steps used in this method are as follows:

---
1. Write the system in matrix form, $AX = B$, where A is the coefficient matrix.

2. Find the inverse A^{-1}.

3. Multiply the equation by A^{-1} to get $A^{-1}AX = A^{-1}B$, which gives the solutions as $X = A^{-1}B$.

Example: Solve the following system of equations:

$$\begin{aligned} -x + 3y + z &= 1 \\ 2x + 5y &= 3 \\ 3x + y - 2z &= -2 \end{aligned}$$

Step 1: Rewrite the system using matrix multiplication:

$$\begin{bmatrix} -1 & 3 & 1 \\ 2 & 5 & 0 \\ 3 & 1 & -2 \end{bmatrix} \begin{bmatrix} x \\ y \\ z \end{bmatrix} = \begin{bmatrix} 1 \\ 3 \\ -2 \end{bmatrix}$$

Writing the coefficient matrix as A, we get

$$A \begin{bmatrix} x \\ y \\ z \end{bmatrix} = \begin{bmatrix} 1 \\ 3 \\ -2 \end{bmatrix}$$

Step 2: Find the inverse of the coefficient matrix A. In this case, the inverse is

$$\begin{bmatrix} -\frac{10}{9} & \frac{7}{9} & -\frac{5}{9} \\ \frac{4}{9} & -\frac{1}{9} & \frac{2}{9} \\ -\frac{13}{9} & \frac{10}{9} & -\frac{11}{9} \end{bmatrix}$$

Step 3: Multiply both sides of the equation (that you wrote in step 1) by the matrix A^{-1}. On the left, you get

$$A^{-1} A \begin{bmatrix} x \\ y \\ z \end{bmatrix} = \begin{bmatrix} x \\ y \\ z \end{bmatrix}$$

On the right, you get

$$A^{-1} \begin{bmatrix} 1 \\ 3 \\ -2 \end{bmatrix} = \begin{bmatrix} -\frac{10}{9} & \frac{7}{9} & -\frac{5}{9} \\ \frac{4}{9} & -\frac{1}{9} & \frac{2}{9} \\ -\frac{13}{9} & \frac{10}{9} & -\frac{11}{9} \end{bmatrix} \begin{bmatrix} 1 \\ 3 \\ -2 \end{bmatrix} = \begin{bmatrix} \frac{21}{9} \\ -\frac{3}{9} \\ \frac{39}{9} \end{bmatrix} = \begin{bmatrix} \frac{7}{3} \\ -\frac{1}{3} \\ \frac{13}{3} \end{bmatrix}$$

So, the solution is

$$\begin{bmatrix} x \\ y \\ z \end{bmatrix} = \begin{bmatrix} \frac{7}{3} \\ -\frac{1}{3} \\ \frac{13}{3} \end{bmatrix}$$

Now go back to Section 8.4, Solving Systems of Equations, and use the inverse matrix method to solve some systems there.

Now practice solving systems of equations in Section 8.4 using the inverse matrix method.

Additional Exercises

1. Solve the system of linear equations using the inverse matrix of the coefficient matrix of the system.

$$x + 4y = 1$$
$$x - y = 4$$

2. Use the inverse matrix method to solve the following system of linear equations.

$$3x + 4y = 1$$
$$2x - 5y = 0$$

3. Determine the coefficient matrix, A, of the following system of equations.

$$5x - y - 3z = -1$$
$$x - 2y - 2z = -1$$
$$9x - 3y - 5z = -5$$

Write the system of equations in matrix form $AX = B$. Determine the inverse matrix of A, A^{-1}. Solve the system using the inverse matrix method.

Module 10

Sequences

10.1 Sequences

> An **infinite sequence**, or just a **sequence**, is a function with domain the set of natural numbers 1, 2, 3, 4, ...

We typically use n instead of x for the domain values to emphasize the fact that these values are restricted to the natural numbers. Choosing a letter, say a, to name the function, we write a_n for the function values instead of the usual function notation $a(n)$. These values are called the **terms** of the sequence. This notation allows us to think of a sequence as an ordered list $a_1, a_2, a_3, a_4, \ldots$, just as we list the domain values 1, 2, 3, 4, ...

An example of a sequence is 1, 2, 3, 4, ... The terms of this sequence are the natural numbers.

Another example of a sequence is 3, 3.1, 3.14, 3.141, 3.1415, 3.14159, ..., whose terms are better and better approximations of π.

It is important to distinguish the terms of the sequence—the individual numbers that appear in the list—from the subscript n, which specifies the position in the list. When we wish to talk about an arbitrary term in the sequence, we often say "the nth term" or the "ith term" to name a_n or a_i.

Sequences Given by Formula

For many of the sequences we study, we are given a **formula** to calculate each term of the sequence. Rather than list the terms of a sequence given this way, we write the formula in braces. For example, given the formula $a_n = 1/n$, we write $\{a_n\}$ or $\{1/n\}$ to represent this sequence. To create the list of terms, we substitute integer values for n, starting with 1 to get the first term, 2 to get the second term, 3 to get the third term, and so on. If we do that, we obtain the sequence 1, 1/2, 1/3, 1/4, 1/5, ...

Example: Find the tenth term of the following sequence:
$$\frac{n-1}{n(n+1)}$$

Solution: Simply substitute 10 for n to get

$$\frac{9}{10(11)} = \frac{9}{110}$$

Guessing the Formula

A more interesting problem is this: If we are given the sequence through listing the terms, can we obtain a formula for the sequence?

This is a difficult problem, even for simple sequences, and often requires luck, experience, and insight. For example, if we see the numbers $1, 2, 3, \ldots$ appear as successive terms of the sequence, we might assume the formula is n. If we see $1, 4, 9, \ldots$ appear as successive terms, we might assume the formula is n^2. What if we see the numbers $9, 16, 25, \ldots$ appear as successive terms? Then we might assume the formula is $(n+2)^2$. But why would we even bother to guess the formula? One reason is that if we were asked for the 230th term, it would be a lot easier if we could just substitute 230 into a formula.

Example: Find the eighth term of the sequence $1/3, 8/5, 27/7, \ldots$.

Solution: Each number in the sequence is a fraction. The first three numerators are $1, 8$, and 27, which are the cubes of $1, 2$, and 3, respectively. So we know the numerator of the nth term, a_n, is n^3. As for the denominators, they are the odd integers starting from 3. A formula for this is $2n + 1$. Thus, $a_n = n^3/(2n+1)$, and to answer the original question,

$$a_8 = \frac{8^3}{2(8)+1} = \frac{512}{17}$$

10.1. SEQUENCES

In the following table, we give a list of formulas for some well-known sequences.

$1, 2, 3, \ldots$	n
$1, 3, 5, \ldots$	$2n - 1$
$2, 4, 6, \ldots$	$2n$
$4, 7, 10, \ldots$	$3n + 1$
$7, 11, 15, \ldots$	$4n + 3$
$1, 4, 9, \ldots$	n^2
$5, 14, 29, \ldots$	$3n^2 + 2$
$1, 8, 27, \ldots$	n^3
$2, 4, 8, 16, \ldots$	2^n
$10, 100, 1000, 10000, \ldots$	10^n
$+1, -1, +1, -1, +1, \ldots$	$(-1)^{n+1}$
$-1, +1, -1, +1, -1, \ldots$	$(-1)^n$

To illustrate how we might use such a table, suppose we are asked to find a formula for the sequence $3, 5/2, 7/4, 9/8, \ldots$. In this example, we see the terms are fractions. The numerators are $3, 5, 7, 9, \ldots$, and the most similar entry from the table is $1, 3, 5, \ldots$, which has the formula $2n - 1$. As indicated above, we probably just need to alter this formula slightly (see where we talked about $(n+2)^2$). We try $2n + 1$, and sure enough this formula fits the pattern $3, 5, 7, 9, \ldots$. The denominators are $1, 2, 4, 8, \ldots$, and the closest entry in the table is $2, 4, 8, \ldots$, which has the formula 2^n. Again, it looks as though we need to modify this formula slightly. The pattern we want starts at 2, so we try 2^{n-1}, and by substituting in different values for n, we see this is indeed the formula we want. Thus,

$$a_n = \frac{2n+1}{2^{n-1}}$$

Try to detect a pattern for the sequence given in this exercise and then create a formula for the sequence. Use Help *for a hint if you have trouble determining the pattern or finding a formula.*

In this applet, you are given four terms of a sequence. Try to obtain the formula for the nth term. Enter your answer, and then press Check . Press New to get another problem.

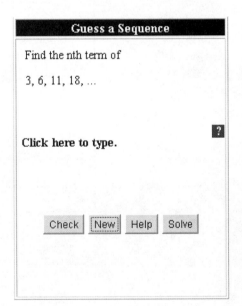

Recursively Defined Sequences

For **recursively defined sequences**, we are given one or more terms of the sequence and a rule as to how to calculate other terms as a formula involving previous terms. As an example, consider the following information: $a_1 = 1$ and $a_n = a_{n-1}/2$. This is a recursively defined

10.1. SEQUENCES

sequence. We are given the first term, namely 1, and we are given the rule that to compute an arbitrary term, we take the term before it and divide it by 2. So what does this sequence look like when we list it? We know that the first term is 1. But what is the second term? Here we must use the recursive formula, which tells us to take the previous term and divide by 2. So, to compute the second term, we take the first term (i.e., 1) and divide by 2 to get 1/2. Now that we have the second term, how do we compute the third? We go back to the recursive formula, which tells us to take the previous term (i.e., the second term), and divide by 2. So, the third term is $(1/2)/2$ or $1/4$. Now compute the fourth term. It is the third term divided by 2, which is $(1/4)/2$ or $1/8$. So, the sequence is $1, 1/2, 1/4, 1/8, \ldots$, which could also have been given by the formula $\{1/2^{n-1}\}$.

Here is a more challenging example.

Example: Define a sequence by $a_1 = 1$, $a_2 = 1$, and $a_n = a_{n-1} + a_{n-2}$. Find a_6.

Solution: Here we are given the first two terms of the sequence and told that any other term is calculated as the sum of the two terms that precede it (the recursive formula). Therefore, to calculate the third term, we add together the second and first terms (which we know). Hence, $a_3 = 1 + 1 = 2$. To calculate the fourth term, we add together the two terms that came before it, namely the third and second term. So, $a_4 = 2 + 1 = 3$. Continuing in this manner, we obtain $a_5 = 3 + 2 = 5$ and $a_6 = 5 + 3 = 8$.

Additional Exercises

1. What is the next term in each of these sequences?

    (a) $1, -1, 1, -1, 1, \ldots$ Next term: _____

    (b) $2, 5, 8, 11, 14, \ldots$ Next term: _____

    (c) $4, -2, 1, -\frac{1}{2}, \frac{1}{4}, \ldots$ Next term: _____

    (d) $-\frac{1}{3}, -\frac{1}{5}, \frac{1}{7}, \frac{1}{9}, \ldots$ Next term: _____

2. Find the eleventh term for each sequence in problem 1. Find a formula for the nth term.

    (a) Eleventh term: _____   (b) Eleventh term: _____

    (c) Eleventh term: _____   (d) Eleventh term: _____

3. Find the third, sixth, and twelfth terms of the sequence $\left\{\dfrac{n}{2n-1}\right\}$.

    Third term: _____   Sixth term: _____   Twelfth term: _____

4. Find the sixth term of the sequence defined by $a_1 = 1$ and $a_n = a_{n-1} + n$.
    Sixth term: _____

5. Find the eighth term of the sequence defined by $a_1 = 1$, $a_2 = 1$, and $a_n = 3a_{n-1} + 2a_{n-2}$.
    Eighth term: _____

10.2 Arithmetic Progressions (AP)

> An **arithmetic sequence** is a sequence in which each term is obtained from the previous term in the sequence by adding a (positive, zero, or negative) constant.

Arithmetic sequences are also known as arithmetic progressions and are usually abbreviated as **AP**.

- The sequence $1, 1, 1, \ldots$ is an arithmetic progression since each term is obtained from the previous term by adding 0.

- The sequence $0, 2, 4, 6, \ldots$ is an arithmetic progression since each term is obtained from the previous term by adding 2.

Of course, there is one term of the arithmetic sequence that is not obtained by this rule, namely the first term. This term is often denoted by a_1. The constant that is added to each term to get the successive term is called the **common difference** and is often symbolized by the letter d. Then for any arithmetic progression, we can list the numbers as $a_1, a_1 + d, a_1 + 2d, a_1 + 3d, \ldots$, or by the usual sequence notation $a_1, a_2, a_3, a_4, \ldots$.

In comparing these two ways to write the same list of numbers, we have a formula for determining each number in the sequence in terms of a_1 and d. Namely,

- $a_2 = a_1 + d$
- $a_3 = a_1 + 2d$
- $a_4 = a_1 + 3d$, etc.

> A typical term of an arithmetic progression is called the nth term and is denoted by a_n. In terms of the first term and common difference, it is given by
>
> $$a_n = a_1 + (n-1)d$$

Some of the sequences given in the table in Section 10.1 are APs. From what we have seen, we now can find the formula for each of those sequences.

Sequence	First Term	Common Difference	Formula for a_n
1, 2, 3, ...	1	1	$1 + (n-1)1 = n$
1, 3, 5, ...	1	2	$1 + (n-1)2 = 2n - 1$
2, 4, 6, ...	2	2	$2 + (n-1)2 = 2n$
7, 11, 15, ...	7	4	$7 + (n-1)4 = 4n + 3$

Here are some more involved examples on using the formula for an AP.

Example: Find the 11th, 31st, and nth term of $7, 13/2, 6, \ldots$.

Solution: The solution to this problem becomes obvious once we recognize the first term and the common difference. The first term is 7. As for the common difference, this is easily found by calculating the difference between the second and first terms, i.e., $13/2 - 7$, which is $-1/2$. So the 11th term is $a_{11} = 7 + 10(-1/2) = 2$. In a similar way, we have $a_{31} = 7 + 30(-1/2) = -8$. The nth term is $a_n = 7 + (n-1)(-\frac{1}{2})$. Now, of course, we would have nightmares leaving that last expression in that form, so we tidy it up and write it as $7 - (n-1)/2$ or even $(15 - n)/2$.

Example: Determine whether the sequence $3, 11, 19, 26, 33, 39, \ldots$ is an arithmetic progression.

Solution: To solve this problem we must ask whether all terms of this sequence are obtained from the previous term by the addition of the (SAME) constant. If it were an arithmetic progression, then the common difference would be $(11 - 3)$ or 8. Certainly, the third term (19) is obtained from the second term (11) by adding 8. However, the fourth term (26), is not obtained by adding 8 to the third term (19). So it is *not* an AP.

In some problems you will be given two terms of an AP, neither of which is the first term, and you will be required to find the first term and common difference. Suppose you are given the values of a_m and a_n where we will assume m is greater than n. We know that

$$a_m = a_1 + (m-1)d$$
$$a_n = a_1 + (n-1)d$$

Subtract the second equation from the first to obtain

$$a_m - a_n = (m-n)d$$

10.2. ARITHMETIC PROGRESSIONS (AP)

We can easily solve this equation for d by dividing both sides by $m - n$ (you should be able to explain why we are not dividing by zero).

> Given the two terms of an arithmetic progression, a_m and a_n with m greater than n, the common difference, d, is given by
> $$d = (a_m - a_n)/(m - n)$$

Once we have found d, we can substitute its value into the equation $a_n = a_1 + (n - 1)d$ (or, if you prefer, $a_m = a_1 + (m - 1)d$) to find the value of the first term, a_1.

Example: Find the common difference and first term of an arithmetic progression if the fifth term is 3 and the seventh term is -4.

Solution: The given information is $a_5 = 3$ and $a_7 = -4$. Then, $d = (a_7 - a_5)/(7 - 5) = (-4 - 3)/(7 - 5) = -7/2$. Substitute this value of d into $a_5 = a_1 + 4d$ to get $3 = a_1 + 4(-7/2)$. Solve for a_1 to see $a_1 = 17$.

Try these next two exercises, which employ the formulas $a_n = a_1 + (n - 1)d$ and $d = (a_m - a_n)/(m - n)$. Practice working with arithmetic sequences until you become efficient at manipulating the formulas.

In this exercise, you are given an AP. Enter the common difference and a formula for the nth term of the AP. Placing the mouse over the question marks will bring up the required formulas.

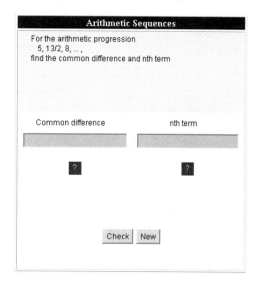

This exercise is a variation on the previous one. You are given two terms of an AP and asked to find the first term and common difference. This involves setting up and solving two linear equations. Enter the common difference and the first term of the AP. Placing the mouse over the question marks will bring up some hints.

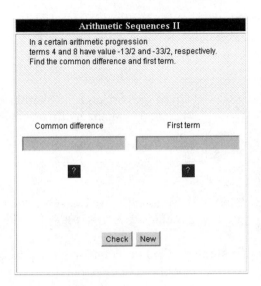

Additional Exercises

1. Which of the following sequences are arithmetic? For those that are, find the common difference d.

    (a) $1, -1, 1, -1, 1, \ldots$

    (b) $2, 5, 8, 11, 14, \ldots$

    (c) $4, 6, 8, 12, \ldots$

    (d) $3, -4, -11, -18, \ldots$

2. Find a formula for the nth term of an arithmetic sequence with first term 3 and common difference 2.

3. Find a formula for the nth term of an arithmetic sequence with first term 3 and common difference -2.

4. The third and seventh terms of an arithmetic sequence are 8 and 21, respectively. Find the common difference and first term.

5. The fifth and eighth terms of an arithmetic sequence are -4 and 11, respectively. Find the common difference and first term.

6. Suppose ten people can sit around a single banquet table with two people on each end. How many can be seated if two such tables are placed end to end? How many can be seated if n such tables are placed end to end?

7. Suppose a square is formed using eight matchsticks. A partial square is then added using six more matchsticks, as shown. How many matchsticks are used if the figure has n partial squares added on?

8. Repeat problem 7, but find the perimeter of the figure formed.

10.3 Sums of Arithmetic Progressions

> Given a sequence $\{a_n\}$, the **sequence of partial sums** $\{S_n\}$ is obtained by adding together the first n terms of $\{a_n\}$. That is, $S_n = a_1 + \cdots + a_n$. Thus, $S_1 = a_1$, $S_2 = a_1 + a_2$, $S_3 = a_1 + a_2 + a_3$, and so on.

For some sequences, including APs, we can find a formula that calculates the values of $\{S_n\}$.

As an example, consider the AP $3, 7, 11, 15, \ldots$. The first partial sum is $S_1 = 3$, and the second partial sum is $S_2 = 3 + 7 = 10$. It would be a lot of work to find the 230th partial sum of this AP. It turns out to be 106,030, but we don't want to have to add 230 terms together to get that answer. Let's devise a shortcut.

First, calculate $S_4 = 3 + 7 + 11 + 15 = 36$. But S_4 is also $15 + 11 + 7 + 3$, since addition is commutative. If we write both versions of S_4 aligned vertically and add, we can see a pattern emerging:

$$\begin{array}{r} 3 + 7 + 11 + 15 = S_4 \\ 15 + 11 + 7 + 3 = S_4 \\ \hline 18 + 18 + 18 + 18 = 2S_4 \end{array}$$

Notice that the same number 18 shows up every time we add vertically. From this we see that $2S_4 = 4(18)$, so $S_4 = 4(18)/2 = 36$.

If we use this technique on S_n instead of S_4, we get

$$\begin{array}{r} 3 + 7 + 11 + \cdots + (3 + (n-1)4) = S_n \\ (3 + (n-1)4) + \cdots + 11 + 7 + 3 = S_n \\ \hline (2(3) + (n-1)4) + \cdots + (2(3) + (n-1)4) = 2S_n \end{array}$$

The same number, this time $2(3) + (n-1)4$, shows up when we add. From this we see that $2S_n = n(2(3) + (n-1)4)$, so

$$S_n = \frac{n(2(3) + (n-1)4)}{2}$$

If we do this one more time using n, a_1, and d, we will have the general formula for S_n.

$$\begin{array}{r} a_1 + \cdots + (a_1 + (n-1)d) = S_n \\ (a_1 + (n-1)d) + \cdots + a_1 = S_n \\ \hline (2a_1 + (n-1)d) + \cdots + (2a_1 + (n-1)d) = 2S_n \end{array}$$

We now see that $2S_n = n(2a_1 + (n-1)d)$.

> The nth partial sum of an arithmetic progression is given by
> $$S_n = \frac{n(2a_1 + (n-1)d)}{2}$$

By noting that $2a_1 + (n-1)d = a_1 + a_n$, we also have the following:

> An alternate formula for the nth partial sum is
> $$S_n = \frac{n(a_1 + a_n)}{2}$$

Example: Find S_{230} for the AP 3, 7, 11, 15, ...

Solution: We substitute $n = 230$, $a_1 = 3$, and $d = 4$ in the formula for S_n to get

$$S_{230} = \frac{230(2)(3) + (229)4}{2} = \frac{230(922)}{2} = 106{,}030$$

You should now do the following exercise, which employs the partial sum formulas. Try to use both versions to get used to evaluating the two forms.

In this exercise, you are given an AP. Find a formula for the required partial sum. Placing the mouse over the question mark will bring up the required formula.

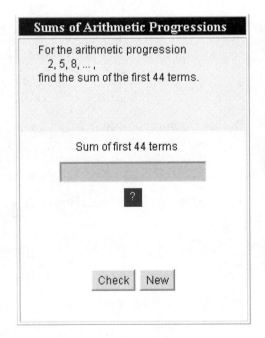

10.3. SUMS OF ARITHMETIC PROGRESSIONS

Additional Exercises

1. Find the eighth partial sum of the AP $3, 5, 7, 9, \ldots$.

2. Find the seventh partial sum of the AP with first term 3 and common difference -2.

3. The third and seventh terms of an AP are 8 and 21, respectively. Find a formula for S_n. Use the formula to find S_5.

4. The fifth and eighth terms of an AP are -4 and 11, respectively. Find a formula for S_n. Use the formula to find S_7.

5. Determine the common difference and first term of an AP given $S_3 = 9$ and $S_5 = 25$.

10.4 Geometric Progressions (GP)

> A **geometric sequence** is a sequence in which each term is obtained from the previous term in the sequence by multiplying by a (positive, zero, or negative) constant.

Geometric sequences are also known as geometric progressions and are usually abbreviated as **GP**.

Here are some examples of geometric sequences:

- The sequence $1, 1, 1, \ldots$ is a geometric progression, since each term is obtained from the previous term by multiplying by 1.

- The sequence $1, 2, 4, 8, \ldots$ is a geometric progression, since each term is obtained from the previous term by multiplying by 2.

Of course, there is one term of the sequence that is not obtained by this rule, namely the first term. This term is often denoted by a_1. The constant by which each term is multiplied to get its successor is called the **common ratio** and is often symbolized by the letter r. Then, for any geometric progression, we can list the numbers as $a_1, a_1 r, a_1 r^2, a_1 r^3, \ldots$, or by the usual sequence notation $a_1, a_2, a_3, a_4, \ldots$.

In comparing these two ways to write the same list of numbers, we have a formula for determining each number in the sequence in terms of a_1 and r. Namely,

- $a_2 = a_1 r$
- $a_3 = a_1 r^2$
- $a_4 = a_1 r^3$, etc.

> A typical (variable) term of the geometric progression is called the nth term and is denoted by a_n. From what we have just seen, it follows that
> $$a_n = a_1 r^{n-1}$$

The following examples show how this formula may be used:

Example: Find the fifth, eighth, and nth term of the geometric progression 4, 1/2, 1/16, ...

Solution: The solution to this problem becomes obvious once we recognize the first term and the common ratio. The first term is 4. As for the common ratio, this is easily found by dividing the second term by the first term, i.e., $(1/2)/4$, which is $1/8$. So the fifth term, $a_5 = 4(1/8)^4 = 1/1024$. In a similar way, we have $a_8 = 4(1/8)^7 = 1/524288$. The nth term is $a_n = 4(1/8)^{n-1}$. We tidy this last expression and write it as $4/8^{(n-1)}$ or even $1/(2 \times 8^{n-2})$. This can also be written as $1/2^{3n-5}$.

Example: Determine whether the sequence 1, 1/2, 1/3, 1/4, ... is a GP.

Solution: Here we are *not* told that this is a GP, rather we are asked to check whether each term of this sequence is obtained from its precursor by multiplication by the SAME constant. If it were a GP, then the common ratio would be $(1/2)/1 = 1/2$. However, the third term $(1/3)$ is not obtained from the second term $(1/2)$ by multiplying by $1/2$. So this is *not* a GP.

In some problems we are given two terms of a GP, neither of which is the first term, and asked to find the first term and common ratio. Suppose we are given terms a_m and a_n of a geometric progression with m greater than n. We have

$$a_m = a_1 r^{m-1}$$
$$a_n = a_1 r^{n-1}$$

Then, dividing a_m by a_n gives

$$a_m/a_n = a_1 r^{m-1}/a_1 r^{n-1} = r^{m-n}$$

From this equation we can solve for r.

Given terms a_m and a_n of a geometric progression, with m greater than n, the common ratio, r, is a solution to

$$r^{m-n} = a_m/a_n$$

If $m - n$ is an odd integer, then

$$r = (a_m/a_n)^{1/(m-n)}$$

If $m - n$ is an even integer, then

$$r = -(a_m/a_n)^{1/(m-n)} \quad \text{or} \quad r = (a_m/a_n)^{1/(m-n)}$$

By substituting the value of r into the equation $a_n = a_1 r^{n-1}$, we can solve to find the value of a_1.

10.4. GEOMETRIC PROGRESSIONS (GP)

Example: Find the common ratio and first term of a GP if the third term is 9 and the fifth term is 4.

Solution: The given information is $a_3 = 9$ and $a_5 = 4$. We have that r is a solution to $r^{5-3} = r^2 = a_5/a_3 = 4/9$. This gives $r = 2/3$ or $-2/3$. Hence, $a_1(4/9) = 9$ or $a_1 = 81/4$.

You should now try this exercise, which utilizes the formula for the nth term of a GP,
$a_n = a_1 r^{n-1}$.

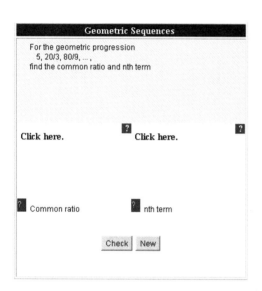

In this exercise, you are given a GP. Enter the common ratio and a formula for the nth term of the GP. For help, place the mouse over the appropriate question mark.

10.4. GEOMETRIC PROGRESSIONS (GP)

Additional Exercises

1. Which of the following sequences are geometric? For those that are, find the common ratio r.

   (a) $1, -1, 1, -1, \ldots$

   (b) $2, 6, 18, 54, \ldots$

   (c) $4, 2, 1, 1/2, \ldots$

   (d) $1/3, -1/6, 1/12, -1/24, \ldots$

   (e) $4, 8/3, 16/9, 32/27, \ldots$

   (f) $8/3, 16/9, 24/15, \ldots$

2. Find a formula for the nth term of a GP with first term 3 and common ratio 2.

3. Find a formula for the nth term of a GP with third term 3 and common ratio $-1/2$.

4. The third and seventh terms of a GP are 1 and 16, respectively. Find the common ratio and first term.

10.5 Sums of Geometric Progressions

Partial Sums and Geometric Progressions

Recall that the sum of the first n terms of a sequence is called the **nth partial sum** of the sequence. As is the case for APs, we can find convenient formulas for the partial sums of geometric sequences. We state the appropriate formulas here first, and then verify and apply each.

> The **nth partial sum** S_n of a geometric sequence with first term a_1 and common ratio r is given by
> $$S_n = na_1 \text{ if } r = 1, \text{ and}$$
> $$S_n = \frac{a_1(1-r^n)}{1-r} \text{ if } r \neq 1$$

First consider the case $r = 1$ for which every a_n is the same as a_1. Then S_n is just a_1 added together n times, so $S_n = na_1$.

Example: Find S_{10} for the GP $4, 4, 4, \ldots$.

Solution: Since $r = 1$, $S_{10} = 10(4) = 40$.

If r is not 1, we can use the following algebraic manipulation to derive the formula for S_n stated above.

Start with the definition of S_n.
$$S_n = a_1 + a_2 + \cdots + a_n$$
Rewrite using the formula for a_n.
$$S_n = a_1 + a_1 r + \cdots + a_1 r^{n-1}$$
Factor out a_1.
$$S_n = a_1(1 + r + \cdots + r^{n-1})$$
Multiply and divide by $1 - r$.
$$S_n = \frac{a_1(1 + r + \cdots + r^{n-1})(1-r)}{1-r}$$
Expand the numerator.
$$S_n = \frac{a_1(1-r^n)}{1-r}$$

Example: Find S_{10} for the GP 5, 10, 20, ...

Solution: Since $r = 2$,
$$S_{10} = \frac{5(1 - 2^{10})}{1 - 2} = \frac{5(-1023)}{-1} = 5115$$

Infinite Sums and Geometric Progressions

If $|r| < 1$ (that is, $-1 < r < 1$), then the values of r^n will approach zero as n increases. The corresponding values of S_n then approach the number
$$S = \frac{a_1(1 - 0)}{1 - r} = \frac{a_1}{1 - r}$$

Note that we are not saying that S_n is ever equal to S, just that it gets closer and closer as n approaches infinity. For any other value of r, the values of S_n don't approach any particular number. Because of this behavior, we make the following definition:

For a GP with first term a_1 and common ratio r, if $|r| < 1$, then we define the **infinite sum** S of all the terms of the GP to be
$$S = \frac{a_1}{1 - r}$$
For all other values of r the infinite sum is not defined.

Example: Find S for the GP 3, 3/5, 3/25, ...

Solution: Since $r = 1/5$, we have $|r| < 1$ and the infinite sum is defined. By the formula above,
$$S = \frac{3}{1 - \frac{1}{5}} = \frac{3}{\frac{4}{5}} = \frac{15}{4}$$

Sometimes the values of S_n are close to S even for small values of n. For instance, the value of S_6 for the GP in the example above is $(15/4)(15{,}624/15{,}625)$, which is very close to $15/4$.

Try the following exercise next to practice using the formulas for S_n and S. Some questions call for numerical answers, and those should be entered as integers or fractions. If the question calls for the formula for S_n, you must type in $(a_1(1 - r^n))/(1 - r)$ with the appropriate values of a_1 and r from the given sequence.

10.5. SUMS OF GEOMETRIC PROGRESSIONS

In this exercise, you are given a GP. Find the partial sum or infinite sum requested. Numerical answers should be entered as integers or fractions. Placing the mouse over the question mark will bring up the required formula.

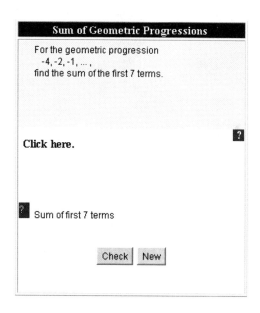

10.5. SUMS OF GEOMETRIC PROGRESSIONS

Additional Exercises

1. Find the sum of the first seven terms of the GP $3, 3/2, 3/4, \ldots$.

2. Find the sum of the first ten terms of the GP $2, -3, 9/2, \ldots$.

3. Find a formula for S_n for a GP with first term 3 and common ratio 2.

4. Find a formula for S_n for a GP with first term 4 and common ratio -5.

5. Find a formula for S_n for a GP with third term 3 and common ratio $-1/2$.

6. Find the infinite sum S for the GP $7, 7/2, 7/4, \ldots$.

7. Find the infinite sum S for the GP $7, -7/2, 7/4, \ldots$.

8. Find S_7 and S for the GP $1, 2/3, 4/9, \ldots$. How close is S_7 to S?

10.6 Repeating Decimals

When we write a decimal number such as 2.3145, we mean

$$2 + \frac{3}{10} + \frac{1}{100} + \frac{4}{1,000} + \frac{5}{10,000}$$

This can be easily converted to rational form by simplifying this sum of rational numbers. It becomes $23,145/10,000$, which can be reduced to $4,629/2,000$.

This example illustrates how every number with a finite number of non-zero digits after the decimal point can be converted to a rational form. We might ask whether the converse is true; that is, does every rational number, when converted to decimal form, have only finitely many non-zero numbers after the decimal point? The answer is no. For example, consider the fraction $1/3$. In order to express $1/3$ in decimal form, divide 1 by 3, and you see that we get $0.3333333\ldots$.

This type of decimal number is called a **repeating** or **recurring** decimal. Other examples of repeating decimals are $2.341555555\ldots$ and $0.3182516251625162516\ldots$. One can view the number 2.3145 as a repeating decimal, namely $2.3145000000\ldots$. With this notion, then,

> Every rational number converted to decimal form is a repeating decimal, and every repeating decimal is a rational number and can be written as a/b where a and b are integers.

(Note that for this discussion, we shall write repeating decimals with the digits that repeat in parentheses. So, we would write the above examples as $0.(3)$, $2.341(5)$, and $0.318(2516)$).

We already know how to convert a rational number into decimal form, but how do we convert a repeating decimal into a rational number form? The answer is provided through the sum of geometric progressions.

In the previous section, we found a formula for the infinite sum S of a GP when $|r| < 1$:

$$S = \frac{a_1}{1-r}$$

Let's see how this formula is used to convert a repeating decimal to a rational form.

Example: Write $0.3(12)$ as the ratio of two integers a and b with no common factor.

Solution: Remember $0.3(12)$ means

$$\frac{3}{10} + \left(\frac{12}{1,000} + \frac{12}{10,0000} + \frac{12}{10,000,000} + \cdots \right)$$

The part in parentheses is a geometric progression with first term 12/1,000 and common ratio 1/100. We can apply the formula above to get the sum of the progression and add that to 3/10.

$$0.3(\overline{12}) = \frac{3}{10} + \frac{\frac{12}{1000}}{1 - \frac{1}{100}} = \frac{3}{10} + \frac{\frac{12}{1000}}{\frac{99}{100}}$$

$$= \frac{3}{10} + \left(\frac{12}{1000}\right)\left(\frac{100}{99}\right) = \frac{3}{10} + \frac{4}{330} = \frac{103}{330}$$

Practice converting between repeating decimal form and rational form in this exercise. The change from decimal to rational form uses the method demonstrated above. The change from rational to decimal form requires you to divide the numerator by the denominator until the repeating block appears.

This applet allows you to convert repeating decimals into their rational form. In the text area on the left, enter the repeating decimal using the parentheses form we discussed. Then press $\boxed{\text{Convert}}$ to see the equivalent rational form.

Additional Exercises

1. Write each of these repeating decimals as fractions.

    (a) $0.5(8) = 0.58888\ldots$

    (b) $6.(39) = 6.393939\ldots$

    (c) $1.23(17) = 1.23171717\ldots$

    (d) $0.(167) = 0.167167167\ldots$

10.7 Amortized Loans

For many loans, a financial institution lends you money and charges you interest on the amount you owe. Each period (usually each month), you pay back some of the amount you borrowed as well as the interest accrued for that period. Usually, the amount you pay each period is constant, and the time of the loan is specified as how long it will take to reduce the amount you owe to zero.

At the beginning of the loan period, almost all of your payment goes toward interest and very little goes toward the principal. Near the end of the loan, the situation is reversed. Typically for home loans, the time of the loan is 30 years, for car loans it is 4 or 5 years, and for home improvement loans 10 to 15 years.

Consider the sequence of numbers $P_0, P_1, P_2, \ldots$, where the values are decreasing. P_0 is the initial amount you borrowed, and P_i is the amount you owe after i periods. Let r be the periodic interest rate, and let x be the amount of each payment. The periodic interest rate is not the stated annual percentage rate (APR), but is that value divided by the number of payment periods in one year. If the APR is 6% and payments are monthly, then the periodic interest rate is $0.06/12 = 0.005$ or $1/2\%$. There is a simple recursive formula for P_i based on the following observation: the balance you still owe after making the current payment P_i is the previous balance P_{i-1} plus the interest charge for the current period minus the periodic payment x. Since the interest charge for the current period is just $P_{i-1}r$, assuming simple interest over the period, we get $P_i = P_{i-1} + P_{i-1}r - x$ or $P_i = P_{i-1}(1+r) - x$ for $i > 0$. Thus,

$$P_1 = P_0(1+r) - x$$

and

$$P_2 = P_1(1+r) - x$$
$$= (P_0(1+r) - x)(1+r) - x = P_0(1+r)^2 - x(1 + (1+r))$$

Continuing in this manner, we find that, in general,

$$P_i = P_0(1+r)^i - x(1 + (1+r) + (1+r)^2 + \cdots + (1+r)^{i-1})$$

Now the expression $1 + (1+r) + (1+r)^2 + \cdots + (1+r)^{i-1}$ is the ith partial sum S_i of the geometric progression with first term 1 and common ratio $1+r$, so,

$$P_i = P_0(1+r)^i - xS_i$$

Now let T be the number of payments over the entire life of the loan. Since on this last payment we want to owe nothing, $P_T = 0$. Therefore,

$$P_T = P_0(1+r)^T - xS_T = 0$$

This allows us to solve for x to get

$$x = \frac{P_0(1+r)^T}{S_T}$$

Now we can use the formula for $S_T = (a_1(1-r^T))/(1-r)$ from the previous section by replacing a_1 with 1 and r with $1+r$ to find

$$S_T = \frac{1-(1+r)^T}{1-(1+r)} = \frac{(1+r)^T - 1}{r}$$

Substituting this into our formula for x gives the amount of each payment as

$$x = \frac{P_0(1+r)^T}{\frac{(1+r)^T-1}{r}} = \frac{rP_0(1+r)^T}{(1+r)^T - 1}$$

Example: Find the monthly payment on a $9000 loan at APR 7% for 10 years.

Solution: Using the formula for x above, we get

$$x = \frac{rP_0(1+r)^T}{(1+r)^T - 1} = \frac{(\frac{0.07}{12})9000(1+\frac{0.07}{12})^{120}}{(1+\frac{0.07}{12})^{120} - 1} \approx \$104.50$$

Use the following exercise to compute monthly payments by entering a loan amount, interest rate, length of loan, and number of payments per year. The reported payment can be checked using the formula above.

The following applet will calculate the amount you must pay each period for an amortized loan. Enter the amount borrowed and the APR, and select the time of the loan and the number of payments per year. Then press the button marked Calculate . On the righthand side of the applet, you will be told your payment as well as the total amount of interest you will pay. On the left, you will see two graphs. The one that appears in red on the web shows how the balance is being reduced, while the graph that appears blue on the web shows the cumulative amount of interest you have paid.

10.7. AMORTIZED LOANS

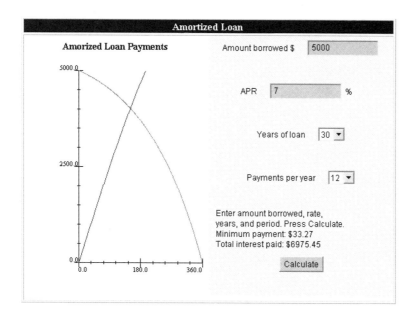

Appendix A

Completing the Square

The method of **completing the square** is a technique used in a variety of problems to change the appearance of quadratic expressions.

The basic idea is to make a perfect binomial square show up by manipulating the original expression. The method is based on the simple observation that, while $x^2 + 10x$ is not a perfect square, $x^2 + 10x + 25$ is. "Completing the square" means putting in the missing 25 that is needed to form the perfect square.

Two obvious concerns are "How do you know you need 25?" and "How can you just add 25 anywhere you want?"

Here are the steps used:

- Start with a quadratic expression like $x^2 + 10x$.
- Square half of the linear coefficient, 10. That would be $(5)^2 = 25$.
- Simultaneously add and subtract 25 and factor the perfect square.
- So $x^2 + 10x = x^2 + 10x + 25 - 25 = (x+5)^2 - 25$.
- Note that the number in the binomial turns out to be half of the linear coefficient.

That's the main idea. Of course there are some variations you'll encounter in different types of problems.

Here are three more examples of completing the square. Make sure you understand the steps involved, especially Example 3 in which the leading coefficient is not 1.

Example 1:

$x^2 + 8x$
$x^2 + 8x + 16 - 16$ since $(\frac{8}{2})^2 = (4)^2 = 16$
$(x+4)^2 - 16$ by factoring $x^2 + 8x + 16$

Example 2:

$x^2 - 3x + 1$
$x^2 - 3x + \frac{9}{4} - \frac{9}{4} + 1$ since $(\frac{-3}{2})^2 = \frac{9}{4}$
$(x - \frac{3}{2})^2 - \frac{5}{4}$ by factoring $x^2 - 3x + \frac{9}{4}$

INTERACTIVE COLLEGE ALGEBRA: A WEB-BASED COURSE © 2005 Key College Publishing

Example 3:

$3x^2 + 5x + 6$
$3(x^2 + \frac{5}{3}x) + 6$ factoring 3 from first two terms
$3(x^2 + \frac{5}{3}x + \frac{25}{36} - \frac{25}{36}) + 6$ since $(\frac{5}{6})^2 = \frac{25}{36}$
$3(x^2 + \frac{5}{3}x + \frac{25}{36}) - \frac{25}{12} + 6$ separating $-\frac{25}{36}$ from the parentheses
$3(x + \frac{5}{6})^2 + \frac{47}{12}$ by factoring $x^2 + \frac{5}{3}x + \frac{25}{36}$

If the expression you are manipulating happens to be part of an equation you are trying to rewrite, you can either do the addition and subtraction as above on one side of the equation, or you have the option of adding the same number to both sides of the equation. Either way, the resulting equation will be equivalent to the one you started with. Here are two additional examples illustrating both approaches followed by a general case. Notice that Example 5 illustrates that completing the square can be applied to the same equation more than once, depending on how many variables appear as squares.

Example 4:

$y = x^2 - 12x + 4$
$y = x^2 - 12x + 36 - 36 + 4$ since $(\frac{-12}{2})^2 = 36$
$y = (x - 6)^2 - 32$ by factoring $x^2 - 12x + 36$

Example 5:

$x^2 + 6x + y^2 - 7y = 3$
$x^2 + 6x + 9 + y^2 - 7y + \frac{49}{4} = 3 + 9 + \frac{49}{4}$ adding 9 and $\frac{49}{4}$ to both sides
$(x + 3)^2 + (y - \frac{7}{2})^2 = \frac{97}{4}$ by factoring both perfect squares

Example 6:

$ax^2 + bx + c$
$a(x^2 + \frac{b}{a}x) + c$ factoring a from first two terms if $a \neq 1$
$a(x^2 + \frac{b}{a}x + \frac{b^2}{4a^2} - \frac{b^2}{4a^2}) + c$ since $(\frac{b}{2a})^2 = \frac{b^2}{4a^2}$
$a(x^2 + \frac{b}{a}x + \frac{b^2}{4a^2}) - \frac{b^2}{4a} + c$ separating $-\frac{b^2}{4a^2}$ from the parentheses
$a(x + \frac{b}{2a})^2 - \frac{b^2 - 4ac}{4a}$ by factoring $x^2 + \frac{b}{a}x + \frac{b^2}{4a^2}$

Appendix B

Broken Wheel Example

This is the solution to the broken wheel example in Section 1.6 (page 37). We restate the problem here:

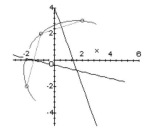

The circular arc in this picture represents the wheel fragment. The three points that have been located are $P_1 = (-2,-2)$, $P_2 = (-1,2)$, and $P_3 = (2,3)$. The line segments connect P_1 to P_2 and P_2 to P_3.

To find the radius, we first find the equations of the perpendicular bisectors of the line segments joining P_1 to P_2 and P_2 to P_3.

	Segment Connecting	
	P_1 to P_2	P_2 to P_3
Slope	$m_1 = \frac{2+2}{-1+2} = 4$	$m_1 = \frac{3-2}{2+1} = \frac{1}{3}$
Perpendicular bisector slope	$m_2 = -\frac{1}{4}$	$m_2 = -3$
Midpoint	$(\frac{-2-1}{2}, \frac{-2+2}{2}) = (-\frac{3}{2}, 0)$	$(\frac{-1+2}{2}, \frac{2+3}{2}) = (\frac{1}{2}, \frac{5}{2})$
Perpendicular bisector	$y - 0 = -\frac{1}{4}(x + \frac{3}{2})$	$y - \frac{5}{2} = -3(x - \frac{1}{2})$
Slope-intercept form	$y = -\frac{1}{4}x - \frac{3}{8}$	$y = -3x + 4$

Now, setting the two formulas for the bisectors equal to each other and solving for x, we get the first coordinate of the center.

$$-\tfrac{1}{4}x - \tfrac{3}{8} = -3x + 4$$
$$3x - \tfrac{1}{4}x = 4 + \tfrac{3}{8}$$
$$\tfrac{11}{4}x = \tfrac{35}{8}$$
$$x = \tfrac{35}{22}$$

Putting the resulting value into one of the bisector equations (we use the second one) gives us the second coordinate of the center.

$$y = -3\left(\frac{35}{22}\right) + 4 = \frac{-105 + 88}{22} = -\frac{17}{22}$$

The center turns out to be the point $(\frac{35}{22}, -\frac{17}{22})$. Lastly, the radius is the distance from the center to any of the original three points. Here, we find the distance using the center and $P_3 = (2, 3)$:

$$r = \sqrt{\left(\frac{35}{22} - 2\right)^2 + \left(-\frac{17}{22} - 3\right)^2} = \sqrt{\frac{81}{484} + \frac{6889}{484}} = \sqrt{\frac{6970}{484}} \approx 3.7948$$

Appendix C

Synthetic Division Example

The first synthetic division demonstration in Section 4.5 steps through the division of $p(x) = x^5 + 4x^3 - 3x^2 + 3x - 1$ by $d(x) = x + 2$. Since $x + 2 = x - (-2)$, the seed is $c = -2$. Also note that the coefficient of x^4 is 0, which is included in the computation. The online demonstration produces the following steps:

$$\begin{array}{r|rrrrrr} -2 & 1 & 0 & 4 & -3 & 3 & -1 \\ \hline & & & & & & \end{array}$$

Write the value of c separated from the coefficients of $p(x)$. Draw a horizontal line.

$$\begin{array}{r|rrrrrr} -2 & 1 & 0 & 4 & -3 & 3 & -1 \\ & \downarrow & & & & & \\ \hline & 1 & & & & & \end{array}$$

Copy the leading coefficient in the first column below the line.

$$\begin{array}{r|rrrrrr} -2 & 1 & 0 & 4 & -3 & 3 & -1 \\ & & -2 \leftarrow & & & & \\ \hline & 1 & & & & & \end{array}$$

Multiply $c = -2$ by 1, and put the result in the second column under the coefficient 0.

$$\begin{array}{r|rrrrrr} -2 & 1 & 0 & 4 & -3 & 3 & -1 \\ & & -2 & & & & \\ \hline & 1 & -2 \leftarrow & & & & \end{array}$$

Add the numbers in the second column, and write the result below the line.

The process of multiplying by c and then adding the numbers in the next column is repeated until the last column is reached. Those two steps are combined in the following:

$$\begin{array}{r|rrrrrr} -2 & 1 & 0 & 4 & -3 & 3 & -1 \\ & & -2 & 4 & & & \\ \hline & 1 & -2 & 8 & & & \end{array}$$

Multiply $c = -2$ by -2, put the result in the next column, add the numbers in the column, and put the result below the line.

$$\begin{array}{r|rrrrrr} -2 & 1 & 0 & 4 & -3 & 3 & -1 \\ & & -2 & 4 & -16 & & \\ \hline & 1 & -2 & 8 & -19 & & \end{array}$$

Multiply $c = -2$ by 8, put the result in the next column, add the numbers in the column, and put the result below the line.

INTERACTIVE COLLEGE ALGEBRA: A WEB-BASED COURSE © 2005 Key College Publishing

APPENDIX C. SYNTHETIC DIVISION EXAMPLE

$$\begin{array}{r|rrrrr} -2 & 1 & 0 & 4 & -3 & 3 & -1 \\ & & -2 & 4 & -16 & 38 & \\ \hline & 1 & -2 & 8 & -19 & 41 & \end{array}$$

Multiply $c = -2$ by -19, put the result in the next column, add the numbers in the column, and put the result below the line.

$$\begin{array}{r|rrrrrr} -2 & 1 & 0 & 4 & -3 & 3 & -1 \\ & & -2 & 4 & -16 & 38 & -82 \\ \hline & 1 & -2 & 8 & -19 & 41 & -83 \end{array}$$

Multiply $c = -2$ by 41, put the result in the next column, add the numbers in the column, and put the result below the line.

The third row (below the line) contains the coefficients of the quotient of the division process followed by the remainder in the final position. Thus the quotient here is $q(x) = x^4 - 2x^3 + 8x^2 - 19x + 41$, and the remainder is $r = -83$.

Appendix D

Answers to Selected Exercises

1.1

1.
2nd	1st
3rd	3rd

3.
I	C	E	D	H
G	J	F	B	A

1.2

1. 4
   15
   $\sqrt{185}$
   $\sqrt{170}$

3. 1, 5

5. $(\frac{5}{2}, 1)$

7. no, $\sqrt{29.25} \approx 5.4$ ft

1.3

1. (a) yes
   (b) no, term of degree 2
   (c) no, terms of degree 2
   (d) no, not an equation
   (e) yes
   (f) yes
   (g) yes

3. (a) yes, $3(2) - 2(-1) = 8$
   (b) no, $3(1) - 2(-1) = 5$
   (c) yes, $3(-4) - 2(-10) = 8$
   (d) no, $3(-2) - 2(-5/2) = -1$

5. (a) $\frac{2}{3}$
   (b) $-\frac{2}{7}$
   (c) $-\frac{2}{7}$
   (d) 0

7. (a) $F = 30 + 10(s - 70)$
       $= 10s - 670$
   (b) If $F = 370$ then $s = 104$ m.p.h.

1.4

1.
(0, 0)	2
(0, −3)	$\sqrt{2}$
(2, −1)	3

3. $(1 \pm \sqrt{3}, -1)$

5. $(x \pm 2)^2 + y^2 = 4$

1.5

1.
3	−2
0	5
7	0
$-\frac{3}{5}$	$\frac{4}{5}$

3. (a) $34 + 27i$
   (b) $-\frac{16}{65} - \frac{43}{65}i$
   (c) $52 - 64i$
   (d) $-\frac{7}{5} + \frac{4}{5}i$

5. $x^2 + 1$

2.1

1. 5, 68

3. $y \geq 0$

5. The y-axis, a vertical line, hits the graph at two points.

7. (a) $T = 14,625 + 0.3(x - 67,700)$
   (b) $25,215

2.2

1. (1) $|x|$, (2) $\frac{1}{2}|x|$, (3) $-2|x|$

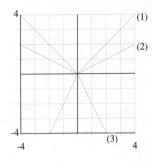

3. (1) x^2, (2) $x^2 + 1$, (3) $x^2 - 3$

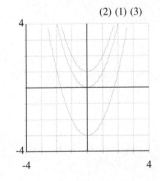

5.

7.

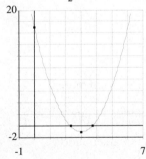

2.3

1. $y = (x - 4)^2 + 11$
3. $(-\frac{3}{2}, \frac{11}{4})$
5. -5

2.4

1. 200 ft × 200 ft, 40,000 ft^2
3. 50 ft × 62.5 ft, 3,125 ft^2

2.5

1. $(3, -1)$, $\frac{6 \pm \sqrt{2}}{2}$

3. $(\frac{3}{4}, -\frac{49}{8})$, $-1, \frac{5}{2}$

APPENDIX D. ANSWERS TO SELECTED EXERCISES

3.1

1. (a) $4x^2 - 1$
   (b) $2x^2 + 1$
   (c) $-2x^2 - 1$
   (d) $3x^4 - 3x^2$
   (e) $\frac{3x^2}{x^2-1}$
   (f) $\frac{x^2-1}{3x^2}$

3. (a) $\frac{20}{3}$
   (b) $\frac{20}{3}$
   (c) 3

5. (a) $\sqrt{x-1} + 1$
   (b) $\sqrt{x}$

7. $(V \circ r)(t) = \frac{256}{3}\pi t^3$

3.2

1. x, x, yes

3. one-to-one

5. not one-to-one

7. $f^{-1}(x) = -\frac{5x-3}{2x}$

9.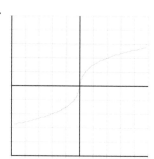

4.1

1. (a) yes, 0, −31, −31
   (b) yes, 3, 1, 8
   (c) yes, 1, 1, 1
   (d) no
   (e) yes, 7, −3, 0
   (f) yes, 1, 1, $\frac{3}{2}$
   (g) no
   (h) no
   (i) yes, 2, 2, 18

3. (a) $f(0) = p(0) = 1$
   (b) $f(0.1) - p(0.1) \approx -0.0000037$
   (c) $f(0.2) - p(0.2) \approx -0.0000549$
   (d) $f(3) - p(3) = -1\frac{1}{16}$
   (e)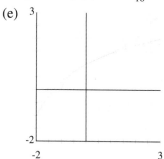

4.3

1. (a) 47
   (b) 15
   (c) $\frac{151}{8}$

3. (a) $-\frac{5}{8}$
   (b) $\frac{9}{8}$
   (c) $\frac{15}{8}$

4.4

1. (a) $-4(x+7)(x-\frac{1}{5})^3(x-9)^2$
   (b) $9x^2(x+\frac{1}{3})(x-\frac{1}{2})(x-5)^4$

4.5

1. (a) $x^3 - x^2 - 6x - 11, -21$
   (b) $x^3 - 4x^2 + 1, 0$
   (c) $x^3 - \frac{5}{2}x^2 - \frac{21}{4}x - \frac{13}{8}, \frac{3}{16}$
   (d) $\frac{1}{2}x^3 - \frac{5}{4}x^2 - \frac{21}{8}x - \frac{13}{16}, \frac{3}{16}$

340 APPENDIX D. ANSWERS TO SELECTED EXERCISES

(e) $\frac{1}{3}x^3 - \frac{11}{9}x^2 - \frac{14}{27}x + \frac{55}{81}, -\frac{29}{81}$

(f) $\frac{1}{2}x^3 - \frac{1}{4}x^2 - \frac{21}{8}x - \frac{97}{16}, -\frac{469}{16}$

4.6

1. (a) $\frac{1}{2}$
   (b) -2
   (c) $-\frac{2}{3}$
   (d) $-\frac{8}{3}$

4.7

1. positive: 2 or 0, negative: 1

3. positive: 1, negative: 2 or 0

5. positive: 2 or 0, negative: 2 or 0

7. positive: 1, negative: 2 or 0

9. positive: 0, negative: 0

4.8

1. (a) 5
   (b) 9
   (c) 4
   (d) $\frac{7}{3}$

3. 4

4.10

1. (a) $\pm(1, 2, 3, 5, 6, 10, 15, 30, \frac{1}{2}, \frac{3}{2}, \frac{5}{2}, \frac{15}{2}, \frac{1}{4}, \frac{3}{4}, \frac{5}{4}, \frac{15}{4})$
   zeros: $-5, -\frac{3}{4}, 2$

   (b) $\pm(1, 2, 4, 8, \frac{1}{3}, \frac{2}{3}, \frac{4}{3}, \frac{8}{3}, \frac{1}{5}, \frac{2}{5}, \frac{4}{5}, \frac{8}{5}, \frac{1}{15}, \frac{2}{15}, \frac{4}{15}, \frac{8}{15})$
   zeros: $-2, -\frac{1}{3}, \frac{4}{5}$

   (c) $\pm(1, 2, 3, 5, 6, 10, 15, 25, 30, 50, 75, 150, \frac{1}{2}, \frac{3}{2}, \frac{5}{2}, \frac{15}{2}, \frac{25}{2}, \frac{75}{2}, \frac{1}{3}, \frac{2}{3}, \frac{5}{3}, \frac{10}{3}, \frac{25}{3}, \frac{50}{3}, \frac{1}{6}, \frac{5}{6}, \frac{25}{6})$
   zeros: $-\frac{5}{3}, \frac{1}{2}, 3, 10$

   (d) $\pm(1, 2, 3, 5, 6, 10, 15, 30, \frac{1}{2}, \frac{3}{2}, \frac{5}{2}, \frac{15}{2}, \frac{1}{4}, \frac{3}{4}, \frac{5}{4}, \frac{15}{4}, \frac{1}{8}, \frac{3}{8}, \frac{5}{8}, \frac{15}{8})$
   zeros: $-\frac{3}{2}, \frac{1}{4}, 2, 5$

4.12

1. (a) $\frac{1}{24}(x+8)(x-3)(x-4)$
   $(0, 4), (-8, 0), (3, 0), (4, 0)$

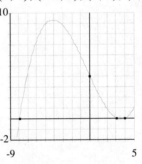

   (b) $\frac{1}{18}(x+3)(5x-6)(x-7)$
   $(0, 7), (-3, 0), (\frac{6}{5}, 0), (7, 0)$

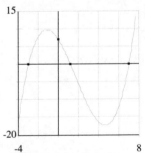

   (c) $\frac{1}{80}(2x+11)(3x+10)(x+2)(x-4)$
   $(0, -11), (-\frac{11}{2}, 0), (-\frac{10}{3}, 0), (-2, 0), (4, 0)$

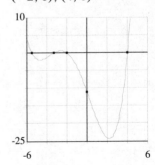

(d) $\frac{1}{3}(x+3)(x+1)(x-2)(x-5)$
$(0, 10), (-3, 0), (-1, 0),$
$(2, 0), (5, 0)$

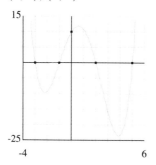

5.1

1. yes
3. yes
5. yes

5.2

1. domain: $x \neq 0$,
   x-intercept: none
3. domain: $x \neq \pm 1$,
   x-intercept: none
5. no removable discontinuity
7. no hole

5.3

1. $f(x) > 0$ for $x < -6$ or $x > 1$
   $f(x) < 0$ for $-6 < x < 1$
3. $f(x) > 0$ for $-4 < x < -\frac{5}{2}$ or $x > 2$
   $f(x) < 0$ for $x < -4$ or $-\frac{5}{2} < x < 2$
5. $f(x) > 0$ for $x > 1$
   $f(x) < 0$ for $x < 1$

5.4

1. (a) ∞
   (b) $-\infty$
3. no vertical asymptote

5. (a) vertical asymptote: $x = -4, 2$
   (b) left of -4: $-\infty$
   (c) right of -4: ∞
   (d) left of 2: $-\infty$
   (e) right of 2: ∞

7. (a) vertical asymptote: $x = 1$
   (b) left of 1: $-\infty$
   (c) right of 1: ∞

5.5

1. (a) below (b) above
3. $y = 1$, above, below

5.6

1. x-intercept: none
   y-intercept: 1
   horizontal asymptote: $y = 0$
   vertical asymptote: $x = 3$

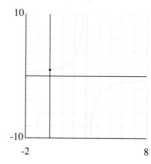

3. x-intercept: none
   y-intercept: -1
   horizontal asymptote: $y = 1$
   vertical asymptote: $x = \pm 1$

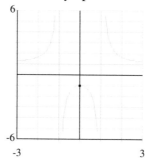

5. x-intercept: 3
   y-intercept: $-\frac{3}{2}$
   horizontal asymptote: $y = 1$
   vertical asymptote: $x = -2$
   hole: $(-1, -4)$

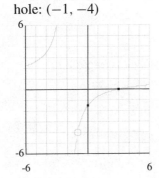

6.1

1. (a) ≈ 1.5 g
   (b) ≈ 210.2 g
   (c) ≈ 111.4 g

3. (a) $\approx \$1269.81$
   (b) $\approx \$1275.48$
   (c) $\approx \$3306.08$
   (d) $\approx \$3306.95$

6.2

1. y-intercept: -1
   horizontal asymptote: $y = -3$

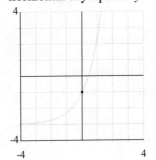

3. y-intercept: 3
   horizontal asymptote: $y = 2$

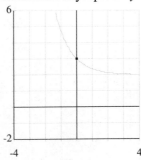

5. y-intercept: 0
   horizontal asymptote: $y = -1$

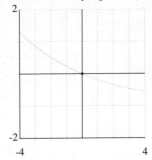

7. y-intercept: $\frac{3}{4}$
   horizontal asymptote: $y = 0$

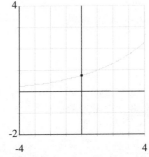

7.1

1. base 3, $x = \log_3 y$

3. (a) $4 = e^z$
   (b) $61 = 3^w$
   (c) $12 = x^5$
   (d) $x = 4^{-3}$
   (e) $z = 3^{-2}$

5. A base cannot be negative.

7.2

1. $\frac{\ln x}{\ln 2}$

3. (a) ≈ 1.0792
   (b) ≈ 2.0792
   (c) ≈ 3.0792
   (d) ≈ 3.7419
   (e) ≈ 5.0875
   (f) ≈ 2.3026

7.3

1. $f^{-1}(x) = \frac{\ln x}{2}$
   domain: $x > 0$

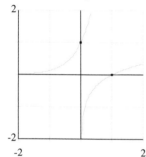

3.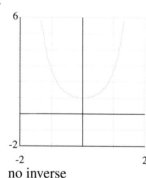
   no inverse

7.4

1. $\log(15)$

3. $\log(\frac{9}{x})$

5. $\ln 3 + 5\ln x + 3\ln z - \ln 2 - 2\ln y$

7.5

1. -1

3. ≈ 2.8074

5. $\frac{\log 15}{\log 3} \approx 2.4650$

7. $\frac{1}{2}$

9. -4.225

11. (a) $T = 25 + 55e^{-kt}$
    (b) $k = -\ln(\frac{7}{55}) \approx 2.06$
    (c) $t \approx 0.63$ hr or ≈ 38 min

13. (a) ≈ 7.95 g
    (b) ≈ 62.5 million years

15. (a) $\$2754.21$
    (b) ≈ 22.9 years

8.1

1. (d)

3. Contains x^2, so not linear. Becomes linear in y and t.

8.2

1. Interchange equations 1 and 2.

3. Equivalent. Interchange equations 1 with equation 1 plus equation 3.

8.3

1. $(7-t, 3+t, t)$, when $t = -1$
   $(x, y, z) = (8, 2, -1)$

3. $(1+t, t, -1-2t)$

8.4

1. $(\frac{17}{5}, -\frac{3}{5})$

3. $(-1, 2, -2)$

5. $(x-1)^2 + (y-1)^2 = 2, r = \sqrt{2}$

7. 9 baskets, 28 animals

9. 12 nickels, 4 dimes, 8 quarters

8.5

1. 2.5 g, 7.5 g

3. 1.8 oz, 0.9 oz, 0.3 oz

8.6

1. $y = \frac{4}{3}x^2 - \frac{7}{3}x + 2$

3. $y = (1-t)x^2 + t$, parameter t

9.1

1. (a) $\begin{bmatrix} -2 & 0 \\ 2 & 2 \end{bmatrix}$

   (b) $\begin{bmatrix} 5 & 1 \\ -5 & -1 \end{bmatrix}$

   (c) $\begin{bmatrix} 5 & 0 \\ 1 & 4 \end{bmatrix}$

   (d) $\begin{bmatrix} 5 & -1 \\ -3 & 7 \end{bmatrix}$

   (e) $\begin{bmatrix} 4 & -5 \\ 6 & 1 \end{bmatrix}$

3. (a) $\begin{bmatrix} 1 & 0 \\ 8 & -1 \\ 1 & 1 \end{bmatrix}$

   (b) not possible

   (c) $\begin{bmatrix} -2 & 2 \\ 16 & -2 \\ -5 & -3 \end{bmatrix}$

   (d) not possible

   (e) not possible

5. $\begin{bmatrix} 4 & -1 & 3 \\ 1 & 2 & -5 \\ 2 & -7 & 0 \end{bmatrix}$,

   $\begin{bmatrix} 4 & -1 & 3 & 4 \\ 1 & 2 & -5 & 1 \\ 2 & -7 & 0 & 8 \end{bmatrix}$

7. $k \times r, k \times (r+1)$

9.2

1. $\begin{bmatrix} 1 & 4 & 1 \\ 0 & 1 & -\frac{3}{5} \end{bmatrix}$, $(\frac{17}{5}, -\frac{3}{5})$

3. $(-1, 2, -2)$

9.3

1. (a) 2

   (b) -2

   (c) -17

3. -2

5. $-2, -6$

7. Interchange two rows, multiply $|A|$ by -1.

   Scale a row by c, multiply $|A|$ by c.

   Adding a multiple of one row to another leaves $|A|$ unchanged.

9.4

1. $(\frac{17}{5}, -\frac{3}{5})$

3. $(-1, 2, -2)$

9.5

1. $A^{-1} = \begin{bmatrix} \frac{1}{2} & \frac{1}{2} \\ -1 & 0 \end{bmatrix}$

   $B^{-1} = \begin{bmatrix} -\frac{1}{2} & \frac{1}{2} \\ 0 & 1 \end{bmatrix}$

   $C^{-1} = \begin{bmatrix} 0 & -\frac{1}{5} \\ \frac{1}{2} & \frac{3}{10} \end{bmatrix}$

9.6

1. $\begin{bmatrix} \frac{17}{5} \\ -\frac{3}{5} \end{bmatrix}$

3. $\begin{bmatrix} -1 \\ 2 \\ -2 \end{bmatrix}$

APPENDIX D. ANSWERS TO SELECTED EXERCISES

10.1

1. (a) -1
   (b) 17
   (c) $-\frac{1}{8}$
   (d) $-\frac{1}{11}$

3. $\frac{3}{5}, \frac{6}{11}, \frac{12}{23}$

5. 2753

10.2

1. (a) not arithmetic
   (b) $d = 3$
   (c) not arithmetic
   (d) $d = -7$

3. $a_n = 3 + (n-1)(-2)$

5. $d = 5, a_1 = -24$

7. $14 + (n-1)6$

10.3

1. 80

3. $S_n = \frac{n(3+(n-1)\frac{13}{4})}{2}, S_5 = 40$

5. $d = 2, a_1 = 1$

10.4

1. (a) -1
   (b) 3
   (c) $\frac{1}{2}$
   (d) $-\frac{1}{2}$
   (e) $\frac{2}{3}$
   (f) not geometric

3. $12(-\frac{1}{2})^{n-1}$

10.5

1. $\frac{381}{64}$

3. $S_n = \frac{3(1-2^n)}{1-2} = -3(1-2^n)$

5. $S_n = \frac{12(1-(-\frac{1}{2})^n)}{1-(-\frac{1}{2})}$
   $= 8(1-(-\frac{1}{2})^n)$

7. $\frac{14}{3}$

10.6

1. (a) $\frac{53}{90}$
   (b) $\frac{211}{33}$
   (c) $\frac{6097}{4950}$
   (d) $\frac{167}{999}$

Index

A
amortized loan, 327–29
arithmetic progression (AP), 301
    common difference, 305
    summing APs, 307

B
base, 181, 195
bounding zeros, 101
Broken Wheel problem, 26, 37, 333–34

C
Cartesian coordinate, 1
center, 25, 37
change of base formula, 202
circle, 25
    center, 25
    radius, 25
common difference, 301
common logarithm, 201
common ratio, 311
Complete Factorization Theorem, 107
completing the square, 331–32
complex conjugate, 32
complex number, 31
    i, 31
    imaginary part, 31
    real part, 31
complex zeros, 141
composition, 75–76
compound interest, 183
conjugate, 141
constant term, 93
continuously compounded interest, 184
coordinate axes, 1
coordinate plane, 1

Cramer's Rule, 279
credit card interest, 184

D
degree, 93
Descartes' Rule of Signs, 125
determinant, 273
distance, 7
distance formula, 8
dividend, 113
Division Algorithm, 102
divisor, 102
domain, 40, 152

E
e, 201–202
elimination method, 241
equivalent system, 233
exponential decay, 223
exponential equation, 219
exponential form, 199
exponential function, 181
exponential growth, 223

F
factor, 94
Factor Theorem, 102
function, 39
    domain, 40
    even, 49
    inverse, 75
    odd, 49
    range, 40
    rational, 135
Fundamental Theorem of Algebra, 107

G

geometric progression (GP), 311
    common ratio, 317
    summing GPs, 317
graph of a function, 42

H

hole, 155, 163, 175
horizontal asymptote, 152
horizontal line test, 84
horizontal shift, 51

I

i, 31
identity matrix, 283
inconsistent system, 228
infinitely many solutions, 228
intercepts, 13, 62–64
Intermediate Value Theorem, 96
inverse function, 83
inverse matrix, 283

L

leading coefficient, 93
leading term, 266
linear equation, 13
line, 13
    parallel, 15
    perpendicular, 15
    slope, 14
loan payment, 327
logarithmic equation, 219
logarithmic form, 196
logarithmic function, 195

M

matrix, 257
    addition, 258
    arithmetic, 258
    augmented, 260
    coefficient, 260
    diagonal entries, 257
    dimensions, 257
    identity, 283
    inverse, 283
    multiplication, 259
    scalar multiplication, 258
    square, 257
    subtraction, 258
midpoint formula, 9
mixture problem, 249
multiplicity, 108

N

natural exponential function, 185
natural logarithm, 201
numerical approximation of
    zeros, 101, 143

O

one-to-one, 83
ordered pair, 1
origin, 1

P

parabola, 61
    vertex, 62
parameter, 236
parametric form, 236
plotting points, 3
point-slope formula, 16
polynomial, 93
    constant term, 93
    degree, 93
    leading coefficient, 93
    zero, 94
population growth, 185
properties of logarithms, 215
Pythagorean Theorem, 8

Q

quadrant, 1
quadratic formula, 64
quadratic function, 61
quotient, 102

INDEX

R
radioactive decay, 183
radius, 25
range, 40
rational function, 149
Rational Zeros Theorem, 135
recursive formula, 297, 327
remainder, 102
Remainder Theorem, 102
repeating decimal, 323
root, 94
row-echelon form, 266
row operation, 265
row-reduction, 265

S
scalar multiplication, 258
scaling, 52
semicircle, 26
sequence, 293
    recursive formula, 297
slope-intercept form, 16
substitution method, 241
synthetic division, 113, 335–36
system of linear equations, 227

U
unique solution, 227

V
vertex, 62
vertical asymptote, 152
vertical line test, 43
vertical shift, 52

X
x-axis, 1
x-intercept, 62–63, 95, 155

Y
y-axis, 1
y-intercept, 62, 64, 96, 175

Z
zero, 94
    complex, 94, 141
    numerical approximation, 101, 143